人生不必
太用力

[德] 埃克哈特·托利 (Eckhart Tolle) 著

李密 译

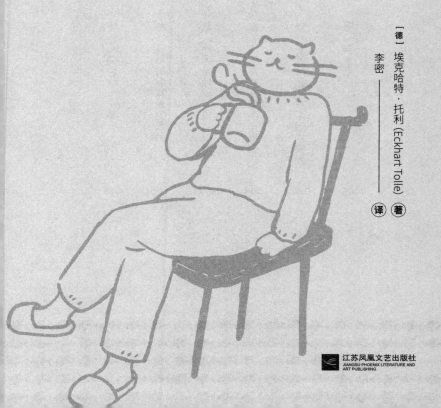

江苏凤凰文艺出版社
JIANGSU PHOENIX LITERATURE AND
ART PUBLISHING

目　录

从内心的宁静开始

当一场风暴来袭时，总会停留一阵子，然后离开。
情绪也一样，来的时候会停留一阵子，然后离开。

（一行禅师）

当你与内在的宁静失去了连接，你也就失去了与自己的连接。当你失去了与自己的连接，你就会迷失在这个世界里。

你内在深处的自我意识，即你是谁，与宁静不可分割。这种"我是……"远比名称或外在形式更为深刻。

宁静是你最真实的本性。那么什么是宁静？宁静是你的内在空间和觉知，在这内在空间里，本页文字被你解读，然后形成思想。没有觉知，便没有解读、没有思想，甚至没有这个世界。

你就是觉知本身，变成人的样子。

✦✦✦

与外部环境的噪音相对应的，是内在思想的嘈杂；而与外部环境的安静相对应的，是内在的宁静。

无论何时，当你身处安静之中，聆听它，留意它，把注意力转向它。聆听安静，可以唤醒你内在的宁静层面，因为只有通过宁静你才能注意到安静。

当你注意到你身处安静之中的那一刻，你并非是在思考。你只是意识到，但并没有在思考。

✦✦✦

当你意识到安静的存在，一种内在宁静的觉醒状态就会立即出现。你完全活在当下。

一树，一花，一草，让你的觉知停留在它们身上，你会发现它们是如此的宁静，如此深植于存在之中。让大自然教导你何为宁静吧。

　　当你看着一棵树，并感觉到它的宁静，你自己也会变得平和起来，你与这棵树在很深的层面上相连接。通过宁静，你感到与你察觉的事物合而为一，无论它是什么。感受到你与万物合而为一，便是爱。

　　安静对找到内在的宁静大有裨益，但并非必要条件。即便身处嘈杂之中，你也能觉知到噪音之下的宁静，觉知到产生噪音的那个空间。那正是你的内在空间，那纯净的觉知、纯净的意识本身。

你会发现，觉知是你所有感觉和思想的基础。意识到觉知本身也就触发了你内心的宁静。

任何干扰性的噪音都可以像安静一样，对我们有所助益。如何做到呢？放下内心对噪音的抗拒，接纳它的真实存在，通过接纳达到内在宁静的国度。

无论何时，无论你以怎样的外在形式呈现，只要你如其所是地全然接纳它，你就会变得宁静、平和。

把你的注意力放在每个间歇，两个念头之间

的空隙，谈话中的静默时间，钢琴或长笛弹奏时音符之间的停顿，又或是你的一呼一吸之间。

当你留心这些间歇，对于"某个事物"的知觉变成了纯粹的觉知，那纯粹的无形的意识从你的内心升起，代替了你对形式的认同。

真正的智慧都是无言的，创造力和解决问题的方法都从宁静中诞生。

宁静代表着没有噪音或者空无一物吗？答案是否定的。宁静就是智慧，是潜藏于深处的意识，任何一个有形之物都诞生于此，你又如何与之分离呢？你所是的这个外在形式正是来源于那里，并且一直被它所滋养。

宁静是万物的本质，大到整个星河，小到花草、树木、鸟儿，以及其他一切有形之物。

　　宁静是这个世界上唯一无形的东西，然而它并非真的是一个东西，也并不属于这个世界。

　　当你静静地看着一棵树或一个人时，到底是谁在看？是远比人更深层的意识在看着它创造的作品。

　　《圣经》里写道，上帝看着自己创造的这个世界，觉得很美好。当你完全心无杂念时，你也能从宁静中看到一个那样的世界。

你需要更多的知识吗？更丰富的信息、更快的计算机或更科学理智的分析能拯救世界吗？难道此时此刻人类最需要的不正是智慧吗？

那么什么是智慧，又如何获得它？智慧往往伴随着安静的能力。只需观看和聆听，无须其他。静静地去看，去听，会激活你内在的智慧，让宁静引导你的一言一行。

开启认知的新格局

人类进化的下一步就是超越思想，

这是我们现阶段的紧急任务。

这并不意味着停止思考，

而是至少不再完全认同于思想，受制于思想。

迷失在思考之中，这就是人类现今的处境。

绝大多数的人，一生都囚禁在自己的思维牢笼里，他们从未超越那受过去制约的、狭隘的、大脑塑造的自我感。

每一个人的内在，都拥有一个比思想更为深邃的意识层面，这便是你的本质。我们可以称之为临在、觉知或无条件的意识。

当你只知道大脑制造的"小我"，并让它主宰你的人生时，你便会将痛苦强加于自己和他人。除非你进入到比思想更为深邃的意识层面，通过无条件的意识，爱、快乐和无限的创造力才会进入你的生命，永恒的内在平和才会进入你的生命。

如果你能认识到，哪怕只是偶尔认识到，你心中出现的想法仅仅是一些想法而已，如果你能看到你的心智、情绪对外界刺激的应激反应模式，那就证明，你已经进入那个比思想更为深邃的意识层面，你的思想和情感皆产生于此，你的人生画卷，将在这个永恒的内在空间中展开。

思想的河流拥有着无穷的动力，很容易就将你卷走。每一个想法都假装自己很重要，它只是想吸引你全部的注意力。

这里有一个新的实践方法：不要太过重视自己的想法。

➤ ➤ ➤

人们是多么容易陷入思维概念的牢笼中啊!

人类的大脑总是渴望知晓、理解和控制，常常把自己的想法和观念误认为是真理。它总说:"道理就是这样的。"你要超越这一观念，进而认识到，无论你怎样理解自己或他人的人生与行为，无论你如何评价任何局面，都只不过是众多观点中的一个观点，众多角度中的一个角度，只不过是一些想法而已。然而，现实是一个统一的整体，万事万物交织在一起，相互影响，没有任何事物可以独立存在。思考会打碎现实，将它切分成零星琐碎的概念碎片。

会思考的大脑是一个强大的工具，但如果你没有察觉到它只是意识的一个很小的层面，只是你的一个很小的层面，而让它完全地掌控你的人生的话，它便会暴露出极大的局限性。

智慧并不是思想的产物，我们深知智慧来源于将自己的全部注意力集中在某个人或某件事这种简单的行为上。关注是原始的智性，是意识本身，它消除了概念化思维形成的障碍，使人们认识到没有任何事物是可以独立存在的。关注是治愈大师，使观察者与被观察者在意识的领域融为一体，使分离得以相融。

　　当你深陷于强迫性思维之中时，便无法认清现实情况，渴望逃离此时此刻。

　　教条是集体性的概念牢笼。不可思议的是，人们欣然接受被囚禁，因为这使他们拥有安全感以及"我知道"的错觉。

没有任何事物能像教条一样给人类带来无尽的痛苦。没错，现实终会发现它的错误，每一个教条早晚都会被打破，然而，除非我们认清其虚伪的本质，否则它将被另一个教条所替代。

那么什么是其虚伪的本质？对思想的认同。

心灵的觉醒是从思考的梦境中觉醒。

意识的领域浩瀚无边，思想根本无法触及。当你不再相信自己所思考的一切，你便跳出了思想，你会清醒地发现这个思考者并不是真正的你。

大脑总是处于"不满足"的状态，总是渴求更多。当你认同了你的大脑，你就很容易觉得无聊和焦躁不安。无聊意味着大脑渴求更多的刺激和更多的精神食粮，却永不知足。

　　当你觉得无聊时，有很多办法可以满足大脑的饥饿感，比如：看杂志、打电话、看电视、上网和购物等。你还可以用一种并不多见的方式，即把精神的空虚与需求转换为身体的需求，用更多的食物暂时性地满足身体。

　　或者你可以让自己置身于无聊焦躁中并观察自己的感受，当你将觉知带入这种感受时，霎时间有一种空间感和宁静感在周围萦绕。起初只是一点点，但随着内在的空间感持续增加，无聊的感觉渐渐消失，变得不再那么强烈和重要了。所以说，就连无聊感都可以教导你是谁，你不是谁。

　　你会发现那个"感到无聊的人"并非是你，

无聊感只是你内部一个受限的能量运动。同样地，气愤、伤感和恐惧也不是你，它们都不属于你，它们只是人类大脑的各种状态，来了又走。

凡是来了又走的，都不是你。

"我感到无聊。"谁知道这个情绪？

"我感到愤怒、伤感、恐惧。"谁又知道这些情绪？

是你，你是这一切的主体，而非那些客体。

任何形式的偏见都意味着你认同了自己的大脑。这表明你再也看不到其他任何人，只是看到你对那个人的看法而已。把一个活生生的人简化成一个看法，这本身可以看作一种暴力。

思考若非植根于觉知，会变得自私自利且功能紊乱。缺乏智慧的聪明充满了危险性和摧毁性，然而这却是大部分人的现状。思想被放大成科学和技术，从本质上讲无所谓好坏，但是，通常诞生这些科学和技术的思想并非植根于觉知之中，致使科学和技术也会变得具有摧毁性。

人类进化的下一步就是超越思想，这是我们现阶段的紧急任务。这并不意味着停止思考，而是至少不再完全认同于思想，受制于思想。

感受你内在的能量，精神上的噪音就会立刻减弱乃至消失。感受能量在你的手、脚、腹腔和胸腔游走，感受你的生命，那个向你体内注入生机活力的生命。

这内在的生命活力，处于波动的情绪和思想之下，身体成了它的出口。

你的内在有一种活力，不只是你的大脑，你的整个生命都能感受到它。在那个存在，你无须思考，你的每一个细胞都充满活力。然而，如果为了达到某些实际的目的需要思考，你便会自然而然地开始思考。当伟大的智慧通过大脑表达自己时，大脑会运作得更为美妙。

　　你或许不曾留意那些"有意识却没在思考"的短暂时刻，它们早已自然而然地出现在你的生命里。当你从事手工活动，或者在房间里走动，又或者在机场柜台等候，你是如此全然临在，致使通常的精神状态减弱，被清醒的存在所替代。抑或你发现自己仰望天空或者聆听某人，不在心里做任何的评价，这时你的觉知变得清澈透明，不被思想所掩盖。

然而，对大脑来说，所有这些都没有意义，也不值得纪念，它有"更重要"的事情去思考，这就是为何这些事情已经发生，你却依然注意不到它们的原因。

　　事实上，能够降临到你身上的最重要的事情莫过于此。这是从思想向存在转变的开端。

　　学会和"未知"的状态和谐相处，这需要你超越大脑。因为大脑害怕"未知"，它总是试图推断和诠释。所以，当你接受"未知"时，你已经超越了大脑，一个更深刻的非概念性的觉知在这种状态里产生。

　　艺术创作、体育运动、舞蹈、教学和咨询等，精通任何领域都意味着思考着的大脑抑或没有牵涉其中，抑或退居二线。此时是一个比你更强大，本质上却又与你同在的智慧在起作用。这已不再是你能掌控的决策程序，而是无意识的正确选择。掌握生命和控制生命是对立的，你和强大的意识密切协作，是意识在行动，在发言，在完成这些工作。

危险时刻能让持续的思考暂时停止，这会让你初步认识到何为临在、觉醒和觉知。

真理总是包罗万象，远非大脑所能理解。思想只是指向真理。比如，"万物在本质上都是一体的"，这只是一个指向，而非解释。充分理解这句话，意味着从你的内在去感受这些文字所指向的真理。

自己是一切幸运的原动力

让我们一起追随内心的光亮，
成为更好的自己。

大脑不停歇地寻找，不只是为了获得精神食粮，还为了它的身份认同，它的自我感。小我由此产生，并不断改造自己。

　　当你想到自己或谈及自己的时候说"我"，通常来说，你是指"我或我的故事"。这个"我"关乎你的喜好和厌恶，恐惧和渴望，这个"我"永不知足。这只是大脑臆想的你，受制于过去并试图在未来得以满足。

　　难道你看不出，这个"我"只是个像浪花一样转瞬即逝的存在。

谁能看穿这一切？谁能意识到身体和精神的形式是短暂的？是"我"，是和过去与未来无关的深层次的"我"。

这困难重重的一生，恐惧和渴望占据了你大部分的精力，到头来还剩下些什么？一个刻在你墓碑上的出生日和死亡日之间，一两英寸长的破折号而已。

对于小我来说，这是个令人沮丧的想法，而对于你，这却是解脱。

若每一种想法都能完全占据你的注意力，说明你认同了大脑里的声音，这些想法就被赋予了自我感，小我便产生了，它是大脑创造的"我"。

大脑创造的小我残缺不全，飘摇不定，于是恐惧和渴望成了它的主导情绪和动力。

当你发现你的大脑里有一个声音假装是你，唠唠叨叨没完没了，你就从对思考无意识的认同中觉醒了。当你觉察到那个声音，你就会认识到那个声音并不是你，那个思考者也不是你，你是认识到这一切的那个人。

认识到自己是那个声音背后的觉知，即为自由。

小我总是在忙于寻找，寻找更多的东西加在自己身上，让自己感觉更完整。这就是小我全神贯注于未来的原因。

当你意识到自己在"为下一刻而活"，你就已经跳出了小我的思维模式，全神贯注于当下的可能性也随之而来。

通过把所有的注意力集中于当下，远比小我思维伟大得多的智慧就进入了你的生命。

当你活在小我里，你总是把当下的时刻贬低为达到目的的手段，你只为未来而活。然而，当你完成了你的目标，却并不觉得满足，至少，你很快就会不满足。

当你更多地关注做某件事的过程，而非这件事所能带来的结果时，你就打破了小我陈旧的制约模式。你的所作所为不仅会变得更加有效，也充满了无尽的满足和愉悦感。

几乎每个小我都至少包含一个我们可以称之为"受害者身份"的元素。一些人给自己塑造的受害者形象是如此强大，以致使其成为小我的核心成分，愤恨和抱怨形成了他们自我感的基本组成部分。

即使你的抱怨"合情合理"，你依旧给自己建立了一个如同牢笼一样的身份，组成这个牢笼的栅栏即思想。看看你对自己做了什么，你的大脑对你做了什么。感受一下你对悲惨故事的情感依赖，回想一下你在想起或讲述它时不由自主的冲动感，像现场目击者一样出现在你的内在状态，你无须做任何事，伴随觉知到来的是转变和自由。

抱怨和抵抗是深受小我喜爱的思维模式，小我通过它们加强自己。对很多人来说，他们的精神情感活动大部分由抱怨和抵抗组成。通过这两种方法，他们使其他人或某种情况变成"错的"，而自己变成"对的"。自己是"对的"让你产生一种优越感，进而加强了你的自我感。当然，事实上，你只是加强了小我的幻觉而已。

你能够察觉到自己内在的这些模式，并认清你脑海里抱怨的声音是什么吗？

小我的自我感需要矛盾冲突，因为通过和一切事物对抗，通过表明这是"我"，而那不是"我"，它的独特身份才能得以加强。

你是否在与人交往时察觉到一些微妙的感受，一些要么比别人优越，要么比别人卑微的感受？那一刻，你正在察觉小我，而它活在比较之中。

嫉妒是小我的副产品，如果好事发生在其他人身上，或者其他人拥有的、知道的、能做的比你多，小我就会感觉被贬低了。小我的身份依赖于比较，并需要确定自己得到的比别人更多，它试图攫取一切。当这些都失败了，它就把一切都归咎于世事不公，或者没有别人幸运，来加强你虚幻的自我感。你的自我感来自怎样的故事，怎样的虚构情节呢？

小我独特的构造需要反对、抵抗和排斥这些情绪，以维持其赖以生存的分离感。所以才会有"我"对抗"他人"，"我们"对抗"他们"。

小我需要与某人或某事产生冲突，这就是为何你渴望平和、快乐和爱，却又无法与它们长期相处；你常说你想要快乐，却总是与烦恼纠缠不清。

从根本上来说，你的烦恼并非来自你的生活处境，而是来自你的思维模式。

你是否对过去所做的或者没有做成功的一些事充满内疚感？当时你肯定是按照自己的意识水平，或者无意识地去做这件事的。如果你那时更加觉醒，更加理智，你或许会做出不一样的选择。

　　内疚是小我构建一个身份、一种自我感的另一种尝试。对于小我来说，自我感是积极的还是消极的并不重要。你在一件事上的成功或失败，只是无意识的表现，是整个人类的无意识。然而，自我却将它个人化，并说"是我做的"，因此你背负了一个"糟糕的"自我形象。

　　如果你设立的小我目标是解放自己、加强自己、提升自己的重要性，即使你完成了这些目标，也依旧不会感到满足。

设定目标没什么不对，但你要知道，达到目标并不是最重要的。万物皆源于当下，这就意味着当下并非一种达到目的的手段，你在每一刻所做的事足以令人满意。你不再把当下贬低为达到目的的手段，便是自我意识。

当被问及佛法的深层含义时，大师说："无我，无业障。"

心想事成的秘密

佛陀曾说，生命只存在于此时此刻。

过去已成为往事，未来尚未发生。

我们只活在一个时刻，那就是当下。

从表面上看，当下只是无数片刻中的一个，人生中的每一天都由无数个片刻组成，每个时刻都发生了不同的事情。然而，如果你能观察得更深入一些，难道不是只有一个片刻吗？难道生命不总是此刻吗？

当下这一刻，是你唯一无法逃离的现实，是人生的不变因素。不论任何事情发生，不论你的人生有多少变故，有一件事是确定不变的，那就是当下。

既然无法逃离，为何不迎接它，善待它呢。

和当下成为朋友，无论到哪儿都觉得自在。

在当下感觉不到的自在,从他处也无法觅得。

当下就是当下,一直如此,你能如其所是地接受它吗?

大脑把人生划分为过去、现在和未来,这些终是虚无缥缈的。

过去和未来只是思维模式,是精神概念。过去只能在当下被记起,你所记起的事情也是在当下发生。至于未来,当它到来时,就变成了当下。所以,唯一真实存在的就是当下。

专注于当下并不意味着否定生命中的其他需求,而是认识到什么是主要的,这样你就可以轻松自如地处理其他次要的事情了。专注于当下也不是说"我不需要做其他任何事了，因为只有当下是唯一的存在"。不是这样。先认清什么是最主要的事，进而和当下成为朋友而非仇敌，承认并尊重它。如果当下能够成为你生命的根基、重中之重，你的人生就会在轻松惬意中展开。

对于收拾整理餐具，拟定经营策略，或者规划一次旅行这样的事，什么才是最重要的呢？是做这些事的过程，还是这些事所带来的结果？是此刻，还是未来的某个时刻？

你是否把当下看作是需要克服的障碍？你是否觉得有一个非常重要的未来时刻需要到达？

几乎所有的人，都是这样度过他们的大部分人生的。然而，除非未来成为现在，否则它永远都是水月镜花。这是一个运转不良的生活方式，它引起了内心无休止的紧张与不安。这也是不尊重生命的表现，生命即是当下，且永远是。

➤➤➤

感受你身体里的生命活力，让它带你到当下。

➤➤➤

从根本上来说，你只有对此刻、对当下负责，才能对你的生命负责。因为生命只存在于此。

对此刻负责指的是从内心深处接受当下的"如是本性"，不与它争辩，并与生命和谐一致。

当下就是当下，一如既往，因为它无法成为其他。没有任何事物是独立存在的，这早已被佛教徒所知晓，现在也被物理学家所证实。表象之下，作为宇宙整体的一部分，所有的事物都相互联系。在当下这一刻，宇宙衍生出万事万物。

当你对这一切说"是"，你就和生命本身的力量与智慧紧密结合，唯其如此，你才能成为使世界向积极方向转变的力量。

这里有一个简单却有效的方法：由内而外地接受当下发生的一切。

当你把注意力转移到当下时，你仿佛从思想的梦境中觉醒，从过去和未来的梦境中觉醒。如此澄澈，如此简单，不给问题和矛盾留一丝空间，只有如其所是的当下。

专注于当下，你会发现生命的神圣。当你临

在，你所观察到的一切都是如此地神圣。你越融入当下，就越能感受到"存在"简单而深远的愉悦，以及一切生命的神圣。

很多人混淆了当下与当下发生的事情，二者绝非同一回事。前者是后者发生的空间，比后者更为深刻。

所以，不要把当下发生的事情和当下混为一谈，当下比在它的空间里发生的任何事情都要深刻得多。

当你步入当下，你也就走出了大脑所能承载的内容。思想的流动逐渐放缓，它们不再吸引你所有的注意力，不再完全占有你，思想与思想之间出现裂痕，如此宽广，如此宁静。你开始认识到，自己的浩瀚与深远是思想难以比拟的。

➤➤➤

思想、情绪、感知和你的一切经历组成了你的生命内容。这就意味着，你的自我感是来自于"我的生命"，于是"我的生命"深感满意，至少你是这么认为的。

你一直在忽视这个显而易见的事实：你内在的"我是"与你的生活里发生的事情毫不相关，和内容毫不相关。"我是"只和当下有关。它始终如一，无论老幼、强弱和胜负，"我是"即当下存在的空间，本质上从未改变。它通常被混淆成生活内容，所以，你只能微弱且间接地从生活的内容里体验到"我是"和当下。换句话来说，你的存在感被这个世界上很多的东西，如境遇、思想等掩盖起来，随着时间的推移，当下也逐渐变得模糊了。

所以，你忘记了自己植根于存在，忘记了自己的神圣，迷失在这个世界里。当人类忘记了自己是谁，困惑、愤怒、绝望、暴力和冲突便蜂拥而至。

然而，想要铭记这个真理并找到回归的路异常容易：

我不是我的思想、情感、感知或经历，也不是我生活的内容。我是生命本身，是一切事物发生的空间，我是意识，是当下，是"我是"。

你心中所想的一切，
都是可以实现的理想

是否有一股力量，在引导你的生命呢？
是否有更崇高的力量，在给予你方向呢？

在最深的层面，真正的你和当下不可分割。

在你的人生里，有很多重要的事，但只有一件事绝对重要。

诚然，在世俗的眼中，成功与否很重要，健康与否很重要，教育程度很重要，富裕与否也很重要，它们会影响你的整个人生。

这些事情的确都很重要，然而相对而言，它们不是绝对的重要。有一件事比它们重要得多，那就是找到你的本质。这件事超越了你短暂的个体，超越了个人化的自我感。

改变生活环境并不能帮你找到平和，而在最深的层面认清你是谁却可以。

在这个星球上，所有的痛苦都来自"我"或者"我们"的个人感觉，这掩盖了你的本质。如果你无法认清自己的内在本质，终其一生，你都在制造痛苦，就是这么简单明了。你不知道自己是谁，大脑就创造了一个替代品，代替你美丽神圣的存在并依附那个恐惧且贫瘠的小我。保护和提高那个伪造的自我感成了你最主要的动力。

有时很多常用的表达，甚至语言结构本身都揭示一个真相：人们并不知道自己是谁。你说"他失去了生命"或者"我的生命"，听起来好像生命是一件你能拥有或丢失的东西一样。

事实是你并不拥有生命，你就是生命本身，是万物合一的生命，是遍及整个宇宙的合一的意识。这个意识以一颗石头、一片草叶、一只动物、一个人、一颗星星或一个银河系的短暂形式来体验它自己。你能在内在感受到自己已经理解了这一切吗？你能感受到自己就是这一切吗？

你需要时间来完成生命中的很多事情，比如学习一项技能、建造一座房子、成为一个专家，或泡一杯茶。

然而，对于生命中最根本的事，时间是不起作用的。这件事就是自我觉醒，需要你超越自我，超越你的姓名和肉体，超越你的过往和故事，去认清自己到底是谁。

你无法在过去或未来找到自己，唯一能找到你自己的时刻就是当下。

探索者希望在未来实现自我领悟或得到启迪，那是因为，作为探索者，他们需要时间，需要未来。如果你相信这些，它就真的需要时间，直到你认识到你并不需要时间去成为自己。

　　当你看着一棵树，你了解了这棵树；当你有一个想法或感受，你认识了这个想法和感受；当你有一个喜悦或痛苦的经历，你体验了这种经历。这些只是看起来正确无误、平淡无奇的陈述，但如果你仔细观察，你会发现它们的结构包含了一个微妙却又基本的错觉，当你使用语言时，这个错觉无法避免。思想和语言创造了一个明显的二元论和一个根本不存在的个体。

　　事实是，认识那棵树、那个想法、那个感觉或经验的人，并不是你。这些事物和感受来源，是你的意识或觉知。当你经营自己的人生，你能看清自己就是意识本身吗？你生命的全部内容在这个意识里，一一呈现。

你说："我想了解自己。"你就是主体，你就是认知，是万物得以被了解的意识。意识无法了解意识，意识包含了所有需要被了解的东西，是一切认知的起源。这个"我"，这个主体，无法再把自己变成认知或意识的客体。

所以，你无法变成自己认知的一个客体。当你在心里把自己当成一个客体时，虚假的小我身份便产生了。"这就是我。"你说。于是你就开始和自己建立一种关系，告诉自己和他人你的故事。

一切表象皆存在于意识之中，而你就是意识，当你认清了这点，你就无须再依赖表象，无须再从各种状况、地点和条件中去寻求自我，你解脱了。换句话说，一切发生或未发生的事，都变得不再那么重要，不再那么沉重和严肃。愉悦进入你的生活，你认识到这个世界只是宇宙的一支舞蹈，是各种外在形式的舞蹈，不多不少，如是而已。

当你知道自己到底是谁，永恒的平和感便油然而生，充满生机。你可以称之为喜悦，因为充满生机的平和正是喜悦的特点。这是认识到你就是生命本质的喜悦，你就是在生命拥有形式之前的本质；这是存在的喜悦，你可以以真实的自己存在。

就像水有固态、液态和气态一样，意识也以不同的方式呈现，可以被"固化"成物质，"液化"成大脑和思想，或"气化"成无形又纯粹的意识。纯粹的意识呈现之前是生命，这个生命通过你的双眼观看这个世界，因为你就是意识。认识到这点，你就能在万物之中发现自己。这是一个完全清醒的觉知状态。你不再是一个承载着沉重过去的实体，不再是诠释一切经历的概念屏幕。停止以诠释的方式观察这个世界，方能觉察到事物的真实本质。语言所能表达的无外乎是：有一个警觉的宁静领域，知觉在此产生。经由你，无形的意识觉察到了它自己。

很多人的生命被渴望和恐惧所驱使。渴望是需要给自己增加更多的东西使自己更加完整；恐惧是害怕因为失去什么东西而变得弱小和逊色。这两者掩盖了一个事实：存在无法被给予或夺走。此刻，完美无缺的存在已属于你。

一次只做一件事

一次只做一件事意味着全身心投入到所做的事情当中，
全神贯注于这件事。

无论何时，如果可以，请你"观察"一下你的内在，看看自己是否无意识地制造了许多冲突，你内在与外在的冲突，你的思想和情感与那一刻你所处的外部环境的冲突。你能感受到内在和外在的现实情况相对抗时，是多么的痛苦吗？

　　一旦你认清这点，你也就意识到，现在你可以自由地放弃这徒劳的对抗，放弃这内在的冲突状态。

　　如果用语言表达出每个时刻你内心的真实感受，那么每天你要说多少遍"我不想在这里"？当你被迫无奈地待在你所处的地方，如堵车路段、工作场所、候机大厅或一个人身边，你会是什么感受？

当然，有些地方你的确可以一走了之，甚至这是最合适的选择。然而，大多数情况下，离开并不是一个选项。在上述所有的例子中，"我不想在这里"的感受根本毫无用处，给你和他人徒增不快。

俗话说："既来之，则安之。"换种说法是："身之所在，即心之所向。"接受这点真的很难吗？

你真的需要在心里给每一个感知、每一个经验贴上标签吗？你这几乎状况百出、矛盾不断的人生，你真的要和它建立一个应激反应式的喜欢或不喜欢的关联吗？或者这只是一个根深蒂固却可以被你打破的思维习惯？要打破这一习惯，你

什么都不必做，只需允许这一刻以它真实的面貌
呈现。

🐟 🐟 🐟

这个习性反映的"不"能够加强小我，而"是"
可以削弱它。所以你的形式身份，即小我，无法
在臣服中幸存。

🐟 🐟 🐟

"我有太多的事情要做。"没错，但是你做
的这些事情质量如何？开车去上班，谈客户，操
作电脑，打杂跑腿，你的日常生活就是处理无穷
无尽的琐事。你对自己所做的事投入了多少？你
是心甘情愿还是无可奈何？你的态度而非你付出
了多少努力，决定了你的人生成功与否。努力意
味着紧张和压力，意味着需要在未来达到一定的
目标或者取得一定的成果。

你是否察觉到在你心中，对所做之事有一丝不情愿？这是对生命的否认，因此真正的成功不可能到来。

　　如果你察觉到这些不情愿，你能放下它，并全然投入到所做的事情当中吗？

　　"一次只做一件事。"禅宗大师如此定义禅的本质。

　　一次只做一件事意味着全身心投入到所做的事情当中，全神贯注于这件事。这是一个臣服的行为，也是一个被赋予力量的行为。

　　全然接受现实情况，会把你带到一个更深的层面，在那里，你的内在状态和自我感，都不再依靠大脑或"好"或"坏"的评价。

当你对"如是本性"俯首称"是"时，当你全然接受了这一刻时，你会感到内在有一种平和的空间感。

表面上，你或许依旧会因为阳光灿烂而心花怒放，因为阴雨连绵而愁眉不展，或许依旧会为赢了百万大奖而欢呼雀跃，为倾家荡产而悲痛万分。然而，无论是开心快乐还是难过悲伤，都变得不再那么重要，它们只是你存在的表面上浮过的涟漪。不论外在条件的性质如何变化，你的内在始终平和、泰然。

对现实情况说"是"揭示了你内在的一个更深的层面，它既不依赖外部条件，也不依赖思想和情感不停波动的内在条件。

🐟 🐟 🐟

当你认识到所有经历短暂易逝的本质，认识

到这个世界无法给你任何具有永恒价值的东西时，臣服也就变得异常简单。你依旧会和他人相遇，会卷入各种经历，参加各种活动，但是却没有了小我的欲望和恐惧。换言之，你不再期待某种情形、某个人、某个地方或某件事来满足你，带给你快乐。它们转瞬即逝、并非完美的本质被你全然接受。

令人惊奇的是，当你不再对它们抱以不合理的期望时，每种情形、每个人、每个地方和每件事都变得令人满意，而且更加和谐、更加平和。

当你全然接受当下，当你不再和现实情况争辩，强迫性的思考会逐渐减少，并被警觉的宁静所取代。你完全清醒，然而大脑却不再以任何方式来标注这一刻。这内在的不抵抗状态，带你进入比人类大脑无限伟大的无条件的意识。这浩瀚的智慧通过你传达它自己，并由内而外支持你。

这就是为何当你放下内在的抵抗时，反而时常发现周遭的一切变得更加美好。

我是在说："享受这一刻，快乐起来吗？"没有。

允许当下"如是"的存在，这就足够了。

臣服是对当下的臣服，而非对一个你为了诠释此刻而杜撰的故事，还试图说服自己相信这个故事。

举个例子，你或许身患残疾，无法再行走了。实际情况就是这样。

你的大脑可能正在杜撰一个故事："我的生

命走到这步田地，要在轮椅上度过余生。生活对我太不公平了，我不应该落得如此下场。"

你能否接受这一刻的"如是本性"，不把它和大脑围绕它杜撰的故事相混淆？

当你停止问"这些为什么会发生在我身上"时，臣服便降临了。

即便是在看起来最无法接受、最苦不堪言的情况下，依旧隐藏着深层的善，每一场灾难也都包含着恩典的种子。

现实中有很多情况，没有答案，也无法解释，生命变得没有意义。或者处于困境中的人向你求助，你却无能为力，不知道该做些什么或说些什么。

当你全然接受你的"不知道"，放弃挣扎着用有限的大脑寻求答案，这时伟大的智慧才能通过你运转起来。甚至思想也能因此受益，因为伟大的智慧可以涌入并激发思想。

有时，臣服意味着放弃试图去理解，与"不知道"和谐相处。

你认不认识这样一些人？他们生命的最主要功能似乎就是给自己和他人制造痛苦、传播悲伤。他们的角色是强化可怕的小我意识，强化不臣服的状态。

可以说，臣服是从抵抗到接受，从"不"到"是"的内在转化过程。当你臣服，你的自我感便发生了转变，从对反应或评价的认同转变为包容反应和评价的空间；从对思想和情感的形式认同，转变为活出自己、承认自己，认识到自己就是那浩瀚而无形的意识本身。

任何事情，包括你无法接受的和你抵抗的，只要你能全然接受它，都能带你进入平和状态。

任凭生命如是存在，顺其自然吧。

| 第七章 |

自然的教育是免费的

留意一朵花，它是如此地临在，
如此地臣服于生命。

我们依赖大自然并不只是为了生存，也需要它指引回家的路，指引逃离思维牢笼的路。我们已经迷失在各种行动、思考、回忆和期待之中，迷失在迷宫一样错综复杂的事物中，迷失在这个问题重重的世界里。

我们忘记了岩石、植物和动物依旧铭记的事，忘记了如何存在、如何保持宁静、如何活出自己，忘记了如何安住于生命所在之处——此时、此刻。

当你把你的注意力停留在任何自然的事物上，任何人类未曾干涉的存在上，你就走出了概念化思维的牢笼，并且，在某种程度上进入与存在相连接的状态，在那里，一切自然的事物依旧充满生机与活力。

一块岩石、一棵树或一只动物，关注它并不意味着思考它，你只需观察，把它带入你的意识。

它会把自己一些本质的东西传递给。你可以感受到它是如此宁静，进而，同样的宁静也在你内部缓缓升起。你感受到它是如此深深地安住于存在之中，与周围的一切合而为一，与所在之处合而为一。意识到这点，你也进入内在深处的宁静之中。

当你在大自然中漫步或休憩时，请完全融入其中，以示对这片领域的敬意。保持宁静，去观察，去聆听，看看每只动物，每株植物，是如何完完全全、毫无保留地做它们自己。与人类不同，它们没有把自己分裂，也没有活在自己的大脑创造的自我形象里。所以，它们不需要劳神费心地保护和加强这些形象。鹿即是鹿，水仙花即是水仙花。

自然万物不仅自成一体，也和宇宙整体紧密相连。它们没有认为自己是一个单独的存在，没有为了试图把自己从整个构造中分离出来而宣称："我"和剩余的宇宙。

对大自然冥想，可以让你从那个"我"中解脱出来，那个惹是生非的"我"。

把你的觉知带入大自然纷繁而微妙的声音中：微风拂叶的瑟瑟声、雨滴坠落的嗒嗒声、昆虫的哼唱声以及破晓的鸟鸣声。全神贯注地去聆听，在这些声音之外有一种更伟大的东西，一种思想所无法理解的神圣性。

你没有创造自己的身体，也不能控制自己的身体机能，所有这些，是比人类大脑伟大得多的智慧在起作用，也正是它支撑着整个大自然。除非通过感受你体内的生命活力以及生机勃勃的存在，从而认识到自己内在的能量，否则你根本无法接近它分毫。

一只快乐的狗，追逐嬉闹，无条件地爱着主人，随时准备庆祝生命。而它的主人却正忍受着沮丧、焦虑的煎熬，面对重重难题并陷入无尽的沉思，全然没有活在当下。然而，当下是唯一存在的时间和地点。两者的内在状态形成鲜明的对比。人们不禁感到疑惑：和这样的人生活在一起，这只狗是如何保持心智健全，并无忧无虑的呢？

当你只通过大脑和思想观察自然时，无法感受到它的活力和存在，你只能看到它的形式，却察觉不到形式内神圣的秘密——生命。大自然不是一个商品，用来谋取利润、知识，或者其他功利主义的目的：森林用来采伐木材；鸟儿用来做研究；高山用来开采或征服。

当你观察大自然时，不要被思想和大脑全部占据，留一些空间。以这种方法接近大自然，它会积极地回应你，并参与到人类的进化和宇宙意识的觉醒中来。

留意一朵花，它是如此地临在，如此地臣服于生命。

那株养在家里的植物，你是否真正观察过它？是否让这个熟悉又神秘，被我们称之为植物的存在告诉你它的秘密？是否注意到它们有多么平和？是否留意到它是如何被宁静围绕着？当你感知到这株植物所散发出来的宁静和平和时，它就成了你的心灵导师。

　　观察一只动物，一朵花，一棵树，看看它是如何安住于存在之中。它只是它自己，有着无上的尊贵、纯洁与神圣。然而，若想要看到这些，你必须超越给大自然命名和分类的思维癖好。当你超越了这些癖好，你会感受到大自然妙不可言的层面，这一层面绝非感官所能体会。那是和谐，是神圣，不仅弥漫于整个大自然，也渗透了你。

正如你呼吸的空气一样，你呼吸的这个过程同样也属于大自然。

当你关注自己的呼吸时，你会发现并不是你在呼吸。呼吸是自然而然发生的事。如果你需要记忆提醒自己去呼吸，那很快就会窒息而亡。如果你试图停止呼吸，大自然将不战而胜。

通过全神贯注地觉察自己的呼吸，你和大自然亲密而有力地联结在一起。这是一件既具有治愈作用，又具有极大力量的事情。它带来了一场意识领域的转变，从思维的概念性世界变成内在无条件的意识领域。

你需要大自然作为你的老师，帮助你和存在重新相互连接。然而不仅是你需要大自然，大自然也需要你。

　　你和大自然是不可分割的。我们都是合一生命的组成部分，这个生命通过无穷无尽的形式来呈现自己，这些形式遍及宇宙且完全相互连接。通过一花一树，你认识了何为神圣、美丽、高贵以及妙不可言的宁静，与此同时，你也赋予了这一花一树某些东西。通过你的觉察和觉知，大自然也了解了自己，了解了自己的美和神圣。

　　一个伟大而寂静的空间，将整个大自然拥入它的怀抱，也将你拥入它的怀抱。

　　唯有内在宁静，你方能接近岩石、植物和动物栖息的宁静领域；也唯有嘈杂的思想平息下来，你方能和自然在更深的层面相连接，并超越过度思考产生的分离感。

思考是生命进化的一个阶段。早在思想诞生
之前，大自然就存在于纯粹的宁静中了。一花一
鸟，一木一石，它们对自身的美和神圣一无所知。
当人类变得宁静时，便超越了思想。在超越思想
的宁静里，觉察和觉知得到了升华。

大自然可以带你进入宁静，这是大自然给你
的礼物。当你在宁静的领域里感知并融入大自然，
那个领域便弥漫着你的觉知，而这是你给大自然
的礼物。

大自然通过你觉察到自己的存在。它一直在
等你，可以说，有数百万年之久。

| 第八章 |

他人是地狱，
也可以是天堂

你能给身边的人最棒的礼物之一，
就是你的存在方式。

我们总是很快就可以对一个人形成一种看法，得出一个结论。给他人贴上标签、伪造一个身份、宣布对其正义的评判，这一切着实满足了小我。

每个人都习惯了运用确定的思考和行为方式，这一方面是由遗传基因决定的，另一方面是由他们的童年经历和文化环境决定的。这并不是他们真实的自己，而是他们表现出的自己。

当你宣布对某人的评判时，你混淆了他们和那些受限制的思维模式。你这么做，本身就是一种受限制的、无意识的行为模式。你给他人扣上了一个概念性身份，这个假身份变成一个牢笼，不仅禁锢了他们，也禁锢了你自己。

放弃评判并不意味着对他们的所作所为视若无睹，而是把他们的行为看成一种受限制的外在形式。你看见了，并如其所是地接受了这种形式，无须由此杜撰一个身份给他人。

放弃评判，把你和他人从对条件、形式和大脑的认同中解放出来。小我便无法再操控你的人际关系了。

只要是小我在操控你的人生，你绝大多数的思想、情绪和行动都是源自渴望和恐惧。在人际关系中，你要不就是渴望从他人那里得到好处，要不就是害怕从他人那里招些坏处。

你想从他们那里得到的无非是快乐、利益、认同、赞扬及关注，也或许是通过对比确定你比他们更尊贵、更富有、更聪明，以此来加强自我感。你所害怕的是事实恰好相反，他们会在某些方面削弱你的自我感。

当你专注于当下，而非把它当作达到某种目的的手段，你就超越了小我，超越了不由自主地利用他人谋取利益的冲动，不再以他人为代价来提升自己。当你专注于和你互动的那个人，除非是为了解决实际问题，过去和未来会被你摒弃在这关系之外。当你全然临在于每一个你遇到的人，你就放弃了以给他人杜撰概念性身份的方式，去诠释他们是谁，在过去做过什么。你可以和他人互动，不再受小我的渴望和恐惧这种伎俩的影响。专注，即警觉的宁静，是这一切的关键所在。

超越渴望和恐惧的人际关系，是多么美妙呀。爱，无所欲，亦无所惧。

想要理解一个人的本质，实际上你无须知道关于他们的任何事情，他们的过去、历史和故事，统统都毫无助益。我们混淆了了解和理解，理解是非概念性的认知。了解和理解是完全不同的认知模式，前者涉及形式，由思想运作；后者涉及无形，由宁静运作。

了解对解决实际问题有帮助。在这方面，没有了解我们什么都做不了。然而，当它成了人际关系的主导模式，就变得有局限性，乃至破坏性。思想和概念构建了一个人工的屏障，把人与人分离开来，人们的交流不再植根于存在，而是大脑。只有消除概念上的屏障，爱才能自然而然地出现在所有人际关系中。

大多数的人际交往局限于语言交流，即思想领域。然而，给人际交往加入点宁静的元素，也是必不可少的，尤其是在亲密关系中。

缺乏空间感的人际关系难以持续发展，而空

间感伴随着宁静。去到大自然里吧，一起冥想或者共度一段安静的时光。散步时，坐在车上或待在家里时，一起舒服地沉浸在宁静里。宁静无法被创造，也不必被创造，它已经在那里，只不过时常被精神上的噪音所掩盖罢了，你只需接纳它便好。

如果广阔的宁静消失，人际关系就会被大脑所控制，并被矛盾和冲突轻松取代。倘若有宁静在，它会包容一切。

真正的倾听是另一个把宁静带入人际关系中的方法，当你真正地在倾听某人时，宁静的层面不断蔓延，成为人际关系最根本的部分。真正的倾听是一个稀缺的技能。通常来说，思考占据了一个人大部分的注意力。在与人交谈时，他们最多也只是在琢磨你所说的话，或者准备他们接下

来要说的话。也许他们根本就没有在听，迷失在
自己的思想里。

　　真正的倾听远远超出听觉感知，它是警觉注
意力的出现，是存在空间。语言在这一空间里被
接受，但退居次要位置，它们或许意味深长或许
没有意义。相比你正在倾听的语言，倾听这个行
为本身要重要得多，在你倾听时，意识的存在空
间出现。在那个使觉知相统一的空间里，你和其
他人相遇，没有大脑思考创造的屏障将你们分割。
此刻，其他人不再是"他人"。在那个空间，你
们连接在一起，作为一个意识体、一个觉知体。

　　在和亲朋好友交往时，你是否经历过不断重
复的剧情？那些相对来说无关紧要的分歧，是否
常常引发激烈的争吵和情感伤痛？造成这些的根
本原因，是小我的基本模式：它必须是"对的"，

当然，其他人必须是"错的"，换句话说，就是认同大脑的立场。另一个原因是，小我需要定期和一些事、一些人产生冲突，以加强它赖以生存的与他人之间的分离感。

另外，每个人都背负着在以往的时光里积累的情感伤痛，它们来自过去的个人经历，也来自人类在悠久的历史里积累的集体伤痛。这个痛苦的身体是你内在的一个能量场，它偶尔会取代你，因为它需要经历更多的情感伤痛去供养和补充自己。它试图控制你的思想，并使其变得无比消极。它爱你消极的思想，因为消极的思想与它有相同的振动频率，它便可以从中汲取能量。它也会激发你身边的人的消极情绪，尤其是你的伴侣，以便吞噬由此带来的冲突剧情和情感伤痛。

对痛苦无意识的认同感，制造了如此多的人生不幸，你要如何解脱？直面它的存在，认识到这并不是真正的你，并认清它的本来面目：源于过去的伤痛。当它出现在你或你的伴侣身上时，正视它。当你打破对它无意识的认同感，当你能够从内在察觉它，它便无法再从你那里汲取存活的能量，它将逐渐失去能量供给。

人际交往可以是地狱，也可以是很好的修行。

当你看到一个人，心中泛起浓浓的爱意时；或者当你凝视着大自然的美，感到内在与之呼应时，闭上你的眼睛，感受这种爱和美。它们与真正的你不可分离，它们是你的本性。外在形式只是你内在本质的临时映象，这就是为何外在形式会离你而去，而爱和美却永远与你同在。

你和这个物质世界，和周遭不计其数的事物，以及每天都需要处理的问题究竟是什么关系呢？你坐的椅子和车，你用的笔和茶杯，它们仅仅是达到目的的工具，还是你偶尔也会留意它们，关注它们，承认它们的存在，哪怕仅是一瞬间。当你依赖物质，利用它们提高你在自己眼中或他人眼中的价值，对这些事物的关注就会轻而易举地占据你的整个生命。

　　当你对外在事物有了自我认同，便不会再去欣赏它们的本来面目，因为你是在它们身上寻找你自己。

　　当你欣赏一个物体原本的样子，当你不带精神投射地承认它的存在，你不会不为它的存在心存感激。你或许还能意识到它不是真的没有生命，只是我们的感官感受不到罢了。物理学家会证明这一点，在分子层面，物体的确是一个震动的能量场。

通过忘我地欣赏这个物质领域，你周围的世界会以各种方式，以你的大脑甚至还未开始理解的方式，为你变得生机盎然。

不论你遇到谁，哪怕只是短暂的一瞬间，你是否会把全部的注意力倾注给他，承认他的存在？抑或你仅仅把他们当作达到目的的一种方法、一个功能、一个角色。你与超市收银员、泊车员、修理工和顾客如何交往呢？

片刻的关注足矣。当你注视他们、倾听他们时，警觉的宁静便会出现。或许只有两三秒，或许更长一些，这已足够某些更真实的东西出现，某些比我们通常使用和认同的角色更为真实的东西。

所有的角色都是受制约的意识的一部分，是人类的大脑思维。然而，通过关注产生的东西却是不受任何条件制约的，是隐藏在名字和形式下的本质的你。你的人生不再是依照剧本上演，你成为真实的自己。当这个层面从你的内在出现，它也会从其他人的内在唤起相同的状态。

当然，最终将不再有其他人，你永远和自己相遇。

随处自在，
你便是奇迹

如果你能学会接受，
甚至欢迎出现在你生命里的这些终结，
你或许会发现那种空虚感，那种起初的不安感，
会变成内在广阔深远的平和感。

漫步在没有被人类破坏过的森林，你不但能发现丰富多彩的生命，还能看到倒塌的树木、腐烂的枝叶和分解的物质。环顾四周，生命和死亡正在同时上演。

　　然而，当你仔细观察时，你会发现这些正在分解的树干和腐烂的枝叶，不仅孕育了新生命，它们自己也充满勃勃生机。微生物在工作，分子在重组。所以，死亡无处可寻，有的只是生命形式的转变。你从中领悟到了什么？

　　死亡不是生命的对立面，生命没有对立面。诞生是死亡的对立面，生命是永恒的。

古往今来，圣贤和诗人们早已认识到，人类的存在犹如梦境一般，看起来真实牢固，却又短暂易逝，随时都可能灰飞烟灭。

临终之时，你的人生故事或许真的像一个梦到了终点。然而，即便是在梦里，也一定存在一个真实的本质，一定会有产生梦境的意识，否则，这个梦无法发生。

是我们的肉体创造了这个意识，还是这个意识创造了肉体的梦，创造了某人的梦？

为什么大多数有过濒死经验的人，都不再对死亡感到恐惧？这值得深思。

当然，你知道自己终有一死，但那只是大脑

里的一个概念，直到你第一次"亲自"面对死亡：你或者你身边某人身患重病、发生意外，或者一个深爱之人去世。死亡进入了你的生命，让你意识到自己也难逃一死。

很多人在恐惧中回避死亡，但是当你不再退缩，正视你的身体是短暂的、随时都可能烟消云散这个现实时，会出现一种对你身体和心理的形式，即"我"的非认同感，哪怕很轻微。当你了解并接受所有生命形式的短暂性，非同寻常的平和感便降临于你。

通过正视死亡，你的意识在某种程度上从形式认同里解脱出来。这就是为什么在一些佛教传统里，和尚会定期到陈尸之处，在尸体间打坐冥想。

在西方文化里，依然普遍存在着对死亡的避讳。甚至连老年人也不愿谈及或想起此事，尸体被掩藏起来。一个拒绝死亡的文化，不可避免地会变得流于表面，只关心外在形式。当

死亡被否认，生命便失去了它的深度。认清在
姓名和形式之外我们是谁的可能性，认识一个
超常的维度的可能性，便从我们的生命里消失，
因为死亡恰恰是通往那个维度的大门。

　　人们倾向于对终结感到不安，因为每个终结
都是一场小小的死亡。这就是为什么在很多语言
里，"再见"这个词的意思是"再次相见"。

　　每当一个经历结束，比如朋友聚会、假期或
孩子离开家，你都"死"了一次。随着这个经历
的消失，一个"形式"出现在你的意识里。通常
这个"形式"会留下一种空虚感，而很多人都尽
力不去感受，不去面对。

　　如果你能学会接受，甚至欢迎出现在你生命
里的这些终结，你或许会发现那种空虚感，那种
起初的不安感，会变成内在广阔深远的平和感。

每天通过这种方式体验死亡，你就向生命打开了自己。

➼ ➼ ➼

大多数人觉得他们的身份和自我感异常珍贵，无法割舍。这就是他们对死亡感到如此恐惧的原因。

"我"将不复存在，这听起来令人恐惧、难以想象。其实，你把那个无比珍贵的"我"和你的姓名、形式以及与之相关的故事混淆了。那个"我"只不过是意识领域里一个短暂的形式罢了。

如果你只知道自己的形式身份，就无法认识到，真正弥足珍贵的是你的本质，是你最深处的"我是"，即意识本身。这才是唯一永远与你相随，永远不会离你而去的东西。

每当你的生命发生了重大的损失，例如失去财产、家庭、亲密关系，或者失去名誉、工作、身体机能等，你内在的某些东西就死了。你的自我感会减弱，或许还伴随着一定的迷茫，"失去了这些……我又是谁？"

　　你无意识地把一个形式认同为自己的一部分，这个形式的离开或消失，对你造成极端的痛苦。可以说，它在你的存在构造里留下了空洞。

　　当这样的情况发生时，不要否认或忽视你所感受到的痛苦和悲伤，承认它的存在。谨防你的大脑试图根据你的损失，编造一个以你为主角的悲惨故事，伴随这个角色的是恐惧、气愤和自怨自艾的情绪。进而警惕这些情绪和这个大脑编造的故事后面隐藏的原因，即那个洞，那个空白空间。你可以正视并接受这个奇怪的空虚感吗？如果可以，你会发现它已不再令人恐惧，你或许会惊奇地发现它正散发着平和。

➡ ➡ ➡

人类只是匆匆过客，生命如白驹过隙，有什么是不受生死制约而永恒存在的吗？

设想一下：如果只有一种颜色存在，假定是蓝色，世间万物都是蓝色的，那么蓝色便不存在。除非有不是蓝色的东西，蓝色才能被辨认出来，否则，它就不会"引人注目"，不会存在。

同样的，难道不需要永恒的东西来辨认出那些转瞬即逝的事物吗？换句话说，如果万物，包括你，都是短暂的，那你又如何认识到这短暂性呢？你能发现并目睹所有的形式，包括你都是短暂的这一真相，难道不意味着你拥有某些永恒不朽的东西吗？

在你二十岁的时候，你感到你的身体强壮有

力，而六十岁时，你感到你的身体虚弱衰老，那时，你的思想也与你二十岁时不再相同。但是你的觉知，知道你的身体是年轻还是衰老，知道你的思想已经发生改变的那个觉知从未改变。那个觉知是永恒的，是意识本身，这是无形的合一生命。你会失去它吗？不会，因为你就是它。

在弥留之际，有些人变得极其平和，几乎发光，如同什么东西从正在消失的形式里照射过来。

有时候，一些重病缠身或衰老虚弱的人，在他们人生的最后几周、几个月，甚至几年的时间里，会变得几近透明。当他们注视着你时，你或许能看到他们眼中闪烁的光芒，没有任何精神痛苦。他们已经臣服。所以这个人，这个大脑制造的小我之"我"，已经消失，他们"在身体死亡之前死去"，并找到内在深处的平和，领悟到不死不灭的奥秘。

每一场意外和灾难，都隐藏着潜在的救赎层面，然而我们通常没有意识到它的存在。

死亡的降临给人以出乎意料的重大打击，可以迫使你的意识完全脱离对形式的认同。在你肉体死亡前的最后时刻和死亡的那一刻，你体验到了自己脱离肉体的形式而成为意识。突然，恐惧消失殆尽，只有平和与领悟：一切皆好，死亡只是形式的消失。你会认识到，死亡终究是虚无的，和被你认同成自己的那个形式一样虚无缥缈。

死亡并非像现代文明令你相信的那样无常和恐怖，它是这个世界上最自然的事，和它的对立面"诞生"一样自然，两者密不可分。当你坐在一个濒临死亡的人身边时，时刻提醒自己这一点，这很重要。

当你坐在一个垂死之人的身边，不要抗拒任何这方面的经历。不要否认正在发生的事，不要否认你的感受。当你发现自己什么也不能做时，你会感到无助、难过或愤怒。接受你的感觉，进而向前一步：接受你爱莫能助的事实，完全接受这一点。你无法左右死亡。那么，完全臣服于这个经历的各个方面，臣服于你的感受，臣服于这个垂死之人可能经历的一切痛苦和不安。你臣服的意识状态和随之而来的宁静将大大帮助到这个濒死之人，并缓和他的生死过渡。如果想要说点什么，那些话语就在你内在的宁静里。但此时语言是次要的。

伴随宁静而来的，是莫大的恩赐：平和。

第十章

安渡风暴，从生活中突围

生活中值得嫉妒的人寥若晨星，
但不幸的人比比皆是。
一定的痛苦和烦恼，几乎是人生的必需品。
我们所能做的是接受生活的本来面目，
并且不要为自己制造痛苦，
尽量让生活变得更加有趣和丰富。
也许，人生就此豁然开朗呢？

万物的关联性已经被物理学家所证实。没有任何事物是独立存在的，只是看似独立罢了。我们越是评判它、标注它，就越是孤立它。

生命的整体性被我们的思维切割得支离破碎。然而，也正是生命的整体性造成了这样的局面。这是相互联系的网络，即宇宙的一部分。这一切意味着：万物本来如此，且只能如此。

大多数情况下，我们无法理解一件看起来毫无意义的事在宇宙整体中所扮演的角色。但是，当你认识到这件事在浩瀚的整体中是必不可少的时候，就会开始从内在接受这个实际情况，开始与生命的整体保持协调一致。

如果你能把当下你的感受和经历，完全看作是自己的选择，你就获得了真正的自由，并终结了痛苦。内在和当下和谐一致便是痛苦的终结。

　　痛苦真的是必要的吗？是，也不是。如果你不曾经历那些痛苦，你就没有作为一个人应有的深度、没有谦卑感、没有同情心，你也不会阅读这本书了。痛苦打破了小我的外壳，与此同时，它也就完成了它的任务。

　　所以，痛苦是必要的，直到你意识到它根本无关紧要。

　　痛苦需要大脑为它创造一个"我"，这个"我"

拥有故事，拥有概念性的身份。它需要时间，需要过去和未来。没有时间的痛苦，还剩下些什么？此刻的"如是本性"而已。你或许感到沉重、焦虑、紧张、愤怒甚至憎恶，但这不是痛苦，也不是个人问题。

人类的痛苦并非是属于个人的。它只是你身体某个部位感受到强烈的压力或强大的能量。当你关注它，这个感觉便不会转变成思想，因而也无法唤醒那个痛苦的我。你只需接受一个感觉，看看会发生什么。

对脑海中出现的每一个想法都信以为真，会给你带来无尽的痛苦。现实情况无法使你痛苦，它或许会引起身体的疼痛，但那不是痛苦。是你的思想、你的诠释以及你讲给自己的故事使你感到痛苦。"是我此刻的想法使我感到痛苦。"领悟到这一点，你就打破了对这些想法无意识的认同感。

"真是糟糕的一天！"

"他没回我的电话，真是太没礼貌了！"

"她真让我失望！"

我们通常用抱怨的形式，讲给自己和他人一些小故事，并无意识地把自己变成"对的"，把其他人或事变成"错的"，以此来加强我们匮乏的小我。自己是"对的"使我们置身于想象的优越感中，进而加强了那个虚假的自我感，即小我。与此同时，这个小我也制造了一些敌人，没错，它需要敌人来划定自己的边界，甚至连天气也能服务于这个功能。通过惯性的精神评价和情绪聚焦，你与你生命中出现的人和事建立了个人化的反应关系。这些关系是你自己制造的痛苦，但却没有被当作痛苦对待，因为小我对它们甚为满意。小我需要反应和冲突来巩固自己。倘若没有这些故事，生命将多么简单美好啊。

"下雨了。"

"他没有回电话。"

"我去了，她没有。"

当你遭受痛苦时，当你感到悲伤时，完全临在于当下。悲伤和矛盾无法在当下存活。

当你在心里给某个处境命名或贴上标签，在某种程度上认为它是讨厌的、糟糕的，痛苦便产生了。你痛恨这个处境，这种恨使这个处境个人化，并且唤醒了那个反应的我。命名和标注是积久渐成的习惯，但这一习惯是可以被打破的。从不再给琐碎的事情命名开始练习，如果你错过了航班、打碎了杯子或者滑倒在泥泞里，你可以控制自己，不把这个经历命名为"糟糕的"或"痛苦的"吗？你可以直接接受那个时刻的"如是本性"吗？

称某些事情是糟糕的，会引起情绪聚焦，当你接受它，不再给它命名时，刹那间，无穷的能量便可为你所用。这种聚焦将你与这个能量、这个生命中的能量隔断。

他们吃了分辨善恶树上的果子。停止在心里给一切事物贴上或好或坏的标签，不再区分善恶。当你超越了命名的习惯，宇宙的力量便降临于你。当你和你的经历处于非反应式的关系时，你曾经称之为不好的事物会通过生命本身的力量发生好转，这个转变即使不是立即发生，也会很快到来。从现在开始，停止把某个经历命名为"糟糕的"，从内在接受它，对它说"是"，顺其自然。

　　不论你的生命处境是什么，如果此刻完全接受了它，你会有怎样的感受呢？

　　痛苦有很多或微妙或明显的形式，它们是如此常见，以至于不被视为痛苦，它们甚至还能满足小我。这些形式包括恼怒、烦躁、气愤、对某人某事看不顺眼、憎恨、抱怨。当这些痛苦的形式出现时，学会辨别它们，并且提醒自己：此刻我正在给自己制造痛苦。

　　如果你习惯于给自己制造痛苦，你很可能也在给别人制造痛苦。要结束这些无意识的思维模式异常简单，只需把它们变成有意识的行为，在它们发生的时候察觉到它们。你不可能在有意识的同时，还给自己制造痛苦。

这简直就是奇迹：每个看似"糟糕的"或"邪恶的"人与事的背后，都潜藏着更深的"善"。从内在接受这些实际情况，这个"善"会从内到外把自己完全呈现给你。"不要拒绝恶"是人类的至高真理之一。

一段对话：

"接受这些实际情况。"

"我真的无法做到。我对此感到又焦虑又气愤！"

"那就接受这些。"

"接受我的焦虑和气愤？接受我不愿接受的？"

"是的。接受你不愿接受的事，臣服你不愿臣服的事，看看会发生什么。"

身体上的慢性疼痛是你能遇到的最严厉的人生导师之一。它教导你：抵抗是徒劳的。没有什么比不愿遭受疼痛更合情合理了。然而，如果你可以放弃不愿意，允许疼痛存在，你会发现在内在深处，自己与疼痛已经分离开来，在你和疼痛之间有一个空间，这个空间一直都存在。这需要自觉自愿地去遭受疼痛。当你自觉自愿地承受疼痛，身体的疼痛很快就把你内在的小我燃烧殆尽，因为小我大部分是由抵抗组成的。对于极端的生理缺陷来说，同样如此。

附 录

英文版

STILLNESS SPEAKS

ECKHART TOLLE

CONTENTS |

Chapter 1   Start from Inner Peace

The storm only stops after it lasts
for some time. So does emotions.
(quoted from Thich Nhat Hanh)

When you lose touch with inner stillness, you lose touch with yourself. When you lose touch with yourself, you lose yourself in the world.

Your innermost sense of self, of who you are, is inseparable from stillness. This is the "I Am" that is deeper than name and form.

* * *

Stillness is your essential nature. What is stillness? The inner space or awareness in which the words on this page are being perceived and become thoughts. Without that awareness, there would be no perception, no thoughts, no world.

You are that awareness, disguised as a person.

* * *

The equivalent of external noise is the inner noise of thinking. The equivalent of external silence is inner stillness.

Whenever there is some silence around you — listen to it. That means just notice it. Pay attention to it. Listening to silence awakens the dimension of stillness within yourself, because it is only through stillness that you can be aware of silence.

See that in the moment of noticing the silence around you, you are not thinking. You are aware, but not thinking.

* * *

When you become aware of silence, immediately there is that state of inner still alertness. You are present.

* * *

Look at a tree, a flower, a plant. Let your awareness rest upon it. How still they are, how deeply rooted in Being. Allow nature to teach you stillness.

* * *

When you look at a tree and perceive its stillness, you become still yourself. You connect with it at a very deep level. You feel a oneness with whatever you perceive in and through stillness. Feeling the oneness of yourself with all things is true love.

* * *

Silence is helpful, but you don't need it in order to find stillness. Even when there is noise, you can be aware of the stillness underneath the noise, of the space in which the noise arises. That is the inner space of pure awareness, consciousness itself.

You can become aware of awareness as the background to all your sense perceptions, all your thinking. Becoming aware of awareness is the arising of inner stillness.

* * *

Any disturbing noise can be as helpful as silence. How? By dropping your inner resistance to the noise, by allowing it to be as it is, this acceptance also takes you into that realm of inner peace that is stillness.

Whenever you deeply accept this moment as it is — no matter what form it takes — you are still, you are at peace.

* * *

Pay attention to the gap — the gap between two thoughts, the brief, silent space between words in a conversation, between the notes of a piano or flute, or the gap between the in-breath and out-breath.

When you pay attention to those gaps, awareness of "something" becomes — just awareness. The formless dimension of pure consciousness arises from within you and replaces identification with form.

* * *

True intelligence operates silently. Stillness is where creativity and solutions to problems are found.

* * *

Is stillness just the absence of noise and

content? No, it is intelligence itself — the underlying consciousness out of which every form is born. And how could that be separate from who you are? The form that you think you are came out of that and is being sustained by it.

It is the essence of all galaxies and blades of grass; of all flowers, trees, birds, and all other forms.

* * *

Stillness is the only thing in this world that has no form. But then, it is not really a thing, and it is not of this world.

* * *

When you look at a tree or a human being in stillness, who is looking? Something deeper than the person. Consciousness is looking at its creation.

In the Bible, it says that God created the world and saw that it was good. That is what you see when you look from stillness without thought.

* * *

Do you need more knowledge? Is more information going to save the world, or faster computers, more scientific or intellectual analysis? Is it not wisdom that humanity needs most at this time?

But what is wisdom and where is it to be found? Wisdom comes with the ability to be still. Just look and just listen. No more is needed. Being still, looking, and listening activates the non-conceptual intelligence within you. Let stillness direct your words and actions.

Chapter 2 New Cognition Pattern

Next step of human evolution is to transcend thoughts, which is an urgent task at this stage. Transcending thoughts does not mean quitting thinking, instead, it means human beings should at least not totally identify with and be controlled by their thoughts.

The human condition: Lost in thought.

* * *

Most people spend their entire life imprisoned within the confines of their own thoughts. They never go beyond a narrow, mind-made, personalized sense of self that is conditioned by the past.

In you, as in each human being, there is a dimension of consciousness far deeper than thought. It is the very essence of who you are. We may call it presence, awareness, the unconditioned consciousness.

Finding that dimension frees you and the world from the suffering you inflict on yourself and others when the mind-made "little me" is all you know and runs your life. Love, joy, creative expansion, and lasting inner peace cannot come into your life except through that unconditioned dimension of consciousness.

If you can recognize, even occasionally, the thoughts that go through your mind as just thoughts, if you can witness your own mental-emotional reactive patterns as they happen, then that dimension is already emerging in you as the awareness in which thoughts and emotions happen — the timeless inner space in which the content of your life unfolds.

* * *

The stream of thinking has enormous momentum that can easily drag you along with it. Every thought pretends that it matters so much. It wants to draw your attention in completely.

Here is a new practice for you: don't take your thoughts too seriously.

* * *

How easy it is for people to become trapped in their conceptual prisons.

The human mind, in its desire to know, understand, and control, mistakes its opinions and viewpoints for the truth. It says: this is how it is. You have to be larger than thought to realize that however you interpret "your life" or someone else's life or behavior, however you judge any situation, it is no more than a viewpoint, one of many possible perspectives. It is no more than a bundle of thoughts. But reality is one unified whole, in which all things are interwoven, where nothing exists in and by itself. Thinking fragments reality — it cuts it up into conceptual bits and pieces.

The thinking mind is a useful and powerful tool, but it is also very limiting when it takes over your life completely, when you don't realize that it is only a small aspect of the consciousness that you are.

* * *

Wisdom is not a product of thought. The deep "knowing" that is wisdom arises through the simple act of giving someone or something your full attention. Attention is primordial intelligence, consciousness itself. It dissolves the barriers created by conceptual thought, and with this comes the recognition that nothing exists in and by itself. It joins the perceiver and the perceived in a unifying field of awareness. It is the healer of separation.

* * *

Whenever you are immersed in compulsive thinking, you are avoiding what is. You don't want to be where you are. Here, Now.

* * *

Dogmas are collective conceptual prisons. And the strange thing is that people love their prison cells because they give them a sense of security and a false sense of "I know".

Nothing has inflicted more suffering on humanity than its dogmas. It is true that every dogma crumbles sooner or later, because reality will eventually disclose its falseness; however, unless the basic delusion of it is seen for what it is, it will be replaced by others.

What is this basic delusion? Identification with thought.

* * *

Spiritual awakening is awakening from the dream of thought.

* * *

The realm of consciousness is much vaster than thought can grasp. When you no longer believe everything you think, you step out of thought and see clearly that the thinker is not who you are.

* * *

The mind exists in a state of "not enough" and so is always greedy for more. When you are identified with mind, you get bored and restless very easily. Boredom means the mind is hungry for more stimulus, more food for thought, and its hunger is not being satisfied.

When you feel bored, you can satisfy the mind's hunger by picking up a magazine, making a phone call, switching on the TV, surfing the web, going shopping, or — and this is not uncommon — transferring the mental sense of lack and its need for "more" to the body and satisfy it briefly by ingesting more food.

Or you can stay bored and restless and observe what it feels like to be bored and restless. As you bring awareness to the feeling, there is suddenly some space and stillness around it, as it were. A little at first, but as the sense of inner space grows, the feeling of boredom will begin to diminish in intensity and significance. So even boredom can teach you who you are and who you are not.

You discover that a "bored person" is not who you are. Boredom is simply a conditioned energy movement within you. Neither are you an angry, sad, or fearful person. Boredom, anger, sadness, or fear are not "yours", not personal. They are conditions of the human mind. They come and go.

Nothing that comes and goes is you.

"I am bored." Who knows this?

"I am angry, sad, afraid." Who knows this?

You are the knowing, not the condition that is known.

* * *

Prejudice of any kind implies that you are

identified with the thinking mind. It means you don't see the other human being anymore, but only your own concept of that human being. To reduce the aliveness of another human being to a concept is already a form of violence.

* * *

Thinking that is not rooted in awareness becomes self-serving and dysfunctional. Cleverness devoid of wisdom is extremely dangerous and destructive. That is the current state of most of humanity. The amplification of thought as science and technology, although intrinsically neither good nor bad, has also become destructive because so often the thinking out of which it comes has no roots in awareness.

The next step in human evolution is to transcend thought. This is now our urgent task. It doesn't mean not to think anymore, but simply not to be completely identified with thought, possessed by thought.

* * *

Feel the energy of your inner body. Immediately mental noise slows down or ceases. Feel it in your hands, your feet, your abdomen, your chest. Feel the life that you are, the life that animates the body.

The body then becomes a doorway, so to speak, into a deeper sense of aliveness underneath the fluctuating emotions and underneath your thinking.

* * *

There is an aliveness in you that you can feel with your entire Being, not just in the head. Every cell is alive in that presence in which you don't need to think. Yet, in that state, if thought is required for some practical purpose, it is there. The mind can still operate, and it operates beautifully when the greater intelligence that you "are" uses it and expresses itself through it.

* * *

You may have overlooked that brief periods in which you are "conscious without thought" are already occurring naturally and spontaneously in your life. You may be engaged in some manual activity, or walking across the room, or waiting at the airline counter, and be so completely present that the usual mental static of thought subsides and is replaced by an aware presence. Or you may find yourself looking at the sky or listening to someone without any inner mental commentary. Your perceptions become crystal clear, unclouded by thought.

To the mind, all this is not significant, because it has "more important" things to think about. It is also not memorable, and that's why you may have overlooked that it is already happening.

The truth is that it is the most significant thing that can happen to you. It is the beginning of a shift from thinking to aware presence.

* * *

Become at ease with the state of "not knowing". This takes you beyond mind because the mind is always trying to conclude and interpret. It is afraid of not knowing. So, when you can be at ease with not knowing, you have already gone beyond the mind. A deeper knowing that is non-conceptual then arises out of that state.

* * *

Artistic creation, sports, dance, teaching, counseling — mastery in any field of endeavor implies that the thinking mind is either no longer involved at all or at least is taking second place. A power and intelligence greater than you and yet one with you in essence takes over. There is no decision-making process anymore; spontaneous right action happens, and "you" are not doing it. Mastery of life is the opposite of control. You become aligned with the greater consciousness. "It" acts, speaks, does the works.

* * *

A moment of danger can bring about a temporary cessation of the stream of thinking and thus give you a taste of what it means to be present, alert, aware.

* * *

The Truth is far more all-encompassing than the mind could ever comprehend. Thought can point to the Truth. For example, it can say: "All things are intrinsically one." That is a pointer, not an explanation. Understanding these words means "feeling" deep within you the truth to which they point.

Chapter 3 You Are the Motivity of All Good Fortune

Let inner light leads us to
become better.

The mind is incessantly looking not only for food for thought; it is looking for food for its identity, its sense of self. This is how the ego comes into existence and continuously re-creates itself.

* * *

When you think or speak about yourself, when you say, "I", what you usually refer to is "me and my story". This is the "I" of your likes and dislikes, fears and desires, the "I" that is never satisfied for long. It is a mind-made sense of who you are, conditioned by the past and seeking to find its fulfillment in the future.

Can you see that this "I" is fleeting, a temporary

formation, like a wave pattern on the surface of the water?

Who is it that sees this? Who is it that is "aware" of the fleetingness of your physical and psychological form? I am. This is the deeper "I" that has nothing to do with past and future.

* * *

What will be left of all the fearing and wanting associated with your problematic life situation that every day takes up most of your attention? A dash — one or two inches long, between the date of birth and date of death on your gravestone.

To the egoic self, this is a depressing thought. To you, it is liberating.

* * *

When each thought absorbs your attention completely, it means you identify with the voice in your head. Thought then becomes invested with a sense of self. This is the ego, the mind-made "me". That mentally constructed self feels incomplete and precarious. That's why fearing and wanting are its predominant emotions and motivating forces.

When you recognize that there is a voice in your head that pretends to be you and never stops speaking, you are awakening out of your unconscious identification with the stream of thinking. When you notice that voice, you realize that who you are is not the voice — the thinker — but the one who is aware of it.

Knowing yourself as the awareness behind the voice is freedom.

* * *

The egoic self is always engaged in seeking. It is seeking more of this or that to add to itself, to make itself feel more complete. This explains the ego's compulsive preoccupation with future.

Whenever you become aware of yourself "living for the next moment", you have already stepped out of that egoic mind pattern, and the possibility of choosing to give your full attention to this moment arises simultaneously.

By giving your full attention to this moment, an intelligence far greater than the egoic mind enters your life.

* * *

When you live through the ego, you always

reduce the present moment to a means to an end. You live for the future, and when you achieve your goals, they don't satisfy you, at least not for long.

When you give more attention to the doing than to the future result that you want to achieve through it, you break the old egoic conditioning. Your doing then becomes not only a great deal more effective, but infinitely more fulfilling and joyful.

* * *

Almost every ego contains at least an element of what we might call "victim identity". Some people have such a strong victim image of themselves that it becomes the central core of their ego. Resentment and grievances form an essential part of their sense of self.

Even if your grievances are completely

"justified", you have constructed an identity for yourself that is much like a prison whose bars are made of thought forms. See what you are doing to yourself, or rather what your mind is doing to you. Feel the emotional attachment you have to your victim story and become aware of the compulsion to think or talk about it. Be there as the witnessing presence of your inner state. You don't have to do anything. With the awareness comes transformation and freedom.

* * *

Complaining and reactivity are favorite mind patterns through which the ego strengthens itself. For many people, a large part of their mental-emotional activity consists of complaining and reacting against this or that. By doing this, you make others or a situation "wrong" and yourself "right". Through being "right", you feel superior, and through feeling superior, you strengthen your sense of self. In reality, of course, you are only strengthening the illusion of ego.

Can you observe those patterns within yourself and recognize the complaining voice in your head for what it is?

* * *

The egoic sense of self needs conflict because its sense of a separate identity gets strengthened in fighting against this or that, and in demonstrating that this is "me" and that is not "me".

* * *

In your dealings with people, can you detect

subtle feelings of either superiority or inferiority toward them? You are looking at the ego, which lives through comparison.

Envy is a by-product of the ego, which feels diminished if something good happens to someone else, or someone has more, knows more, or can do more than you. The ego's identity depends on comparison and feeds on "more". It will grasp at anything. If all else fails, you can strengthen your fictitious sense of self through seeing yourself as "more" unfairly treated by life or "more" ill than someone else.

What are the stories, the fictions from which you derive your sense of self?

* * *

Built into the very structure of the egoic self is

a need to oppose, resist, and exclude to maintain the sense of separateness on which its continued survival depends. So there is "me" against the "other", "us" against "them".

The ego needs to be in conflict with something or someone. That explains why you are looking for peace and joy and love but cannot tolerate them for very long. You say you want happiness but are addicted to your unhappiness.

Your unhappiness ultimately arises not from the circumstances of your life but from the conditioning of your mind.

* * *

Do you carry feelings of guilt about something you did—or failed to do—in the past? This much is certain: you acted according to your level of

consciousness or rather unconsciousness at that time.
If you had been more aware, more conscious, you
would have acted differently.

Guilt is another attempt by the ego to create
an identity, a sense of self. To the ego, it doesn't
matter whether that self is positive or negative.
What you did or failed to do was a manifestation of
unconsciousness–human unconsciousness. The ego,
however, personalizes it and says, "I did that", and so
you carry a mental image of yourself as "bad".

* * *

If you set egoic goals for the purpose of freeing
yourself, enhancing yourself or your sense of
importance, even if you achieve them, they will not
satisfy you.

Set goals, but know that the arriving is not all

that important. When anything arises out of presence, it means this moment is not a means to an end: the doing is fulfilling in itself every moment. You are no longer reducing the Now to a means to an end, which is the egoic consciousness.

* * *

"No self, no problem." said the Buddhist master when asked to explain the deeper meaning of Buddhism.

Chapter 4 The Secret of Making All Wishes Come True

The Buddha once said that life only exists at the present moment. Past has been gone and future has not come yet, so it is the present that we only live for.

On the surface it seems that the present moment is only one of many, many moments. Each day of your life appears to consist of thousands of moments where different things happen. Yet if you look more deeply, is there not only one moment, ever? Is life ever not "this moment?"

This one moment— Now—is the only thing you can never escape from. The one constant factor in your life. No matter what happens. No matter how much your life changes, one thing is certain: it's always Now.

Since there is no escape from the Now, why not welcome it, become friendly with it?

* * *

When you make friends with the present moment, you feel at home no matter where you are. When you don't feel at home in the now, no matter where you go, you will carry unease with you.

The present moment is as it is, Always. Can you let it be?

* * *

The division of life into past, present and future is mind made, and, ultimately, illusory.

Past and future are thought forms, mental abstractions. The past can only be remembered Now. What you remember is an event that took place in the Now and you remember it Now. The future, when it comes, is the Now. So the only thing that is real, the only thing that ever is "is" the Now.

* * *

To have your attention in the Now is not a denial of what is needed in your life. It is recognizing what is primary. Then you can deal with what is secondary with great ease. It is not saying, "I'm not dealing with things anymore because there is only the Now." No. Find what is primary first, and make the Now into your friend, not your enemy. Acknowledge it, honor it. When the Now is the foundation and primary focus of your life, then your life unfolds with ease.

* * *

Putting away the dishes, drawing up a business strategy, planning a trip—what is more important: the doing or the result that you want to achieve through the doing? This moment or some future moment?

Do you treat "this moment" as if it were an obstacle to be overcome? Do you feel you have a future moment to get to that is more important?

Almost everyone lives like this most of the time. Since the future never arrives, except "as" the present, it is a dysfunctional way to live. It generates a constant undercurrent of unease, tension, and discontent. It does not honor life, which is Now and never not Now.

* * *

Feel the aliveness within your body. That anchors you in the Now.

* * *

Ultimately you are not taking responsibility for life until you take responsibility for "this moment"— Now. This is because Now is the only place where life can be found.

Taking responsibility for this moment means not to oppose internally the "suchness" of Now, not to argue with what is. It means to be in alignment with life.

The Now is as it is because it cannot be otherwise. What Buddhists have always known, physicists now confirm: there are no isolated things or events. Underneath the surface appearance, all things are interconnected, are part of the totality of the cosmos that has brought about the form that this moment takes.

When you say "yes" to what is, you become aligned with the power and intelligence of Life itself. Only then can you become an agent for positive change in the world.

* * *

A simple but radical practice is to accept whatever arises in the Now—within and without.

* * *

When your attention moves into the Now, there is an alertness. It is as if you were waking up from a dream, the dream of thought, the dream of past and future. Such clarity, such simplicity. No room for problem making. Just this moment as it is.

* * *

The moment you enter the Now with your attention, you realize that life is sacred. There is a sacredness to everything you perceive when you are present. The more you live in the Now, the more you sense the simple yet profound joy of Being and the sacredness of all life.

* * *

Most people confuse the Now with "what happens" in the Now, but that's not what it is. The Now is deeper than what happens in it. It is the space in which it happens.

So do not confuse the content of this moment with the Now. The Now is deeper than any content that arises in it.

* * *

When you step into the Now, you step out of the content of your mind. The incessant stream of thinking slows down. Thoughts don't absorb all your attention anymore, don't draw you in totally. Gaps arise in between thoughts—spaciousness, stillness. You begin to realize how much vaster and deeper you are than your thoughts.

* * *

Thoughts, emotions, sense perceptions, and whatever you experience make up the content of your life. "My life" is what you derive your sense of self from, and "my life" is content, or so you believe.

You continuously overlook the most obvious fact: your innermost sense of "I Am" has nothing to do with what "happens" in your life, nothing to do with content. That sense of "I Am" is one with the Now. It always remains the same. In childhood and old age, in health or sickness, in success or failure, the "I Am"—the space of Now—remains unchanged at its deepest level. It usually gets confused with content, and so you experience "I Am" or the Now only faintly and indirectly, through the content of your life. In other words: your sense of Being becomes obscured by circumstances, your stream of thinking, and the many things of this world. The Now becomes obscured by time.

And so you forget your rootedness in Being, your divine reality, and lose yourself in the world. Confusion, anger, depression, violence, and conflict arise when humans forget who they are.

Yet how easy it is to remember the truth and thus return home:

I am not my thoughts, emotions, sense perceptions, and experiences. I am not the content of my life. I am Life. I am the space in which all things happen. I am consciousness. I am the Now. I Am.

Chapter 5   All Your Ideas Can Become True

Is there a strength that leads your life? Is there a more sublime strength that directs your road?

The Now is inseparable from who you are at the deepest level.

* * *

Many things in your life matter but only one thing matters absolutely.

It matters whether you succeed or fail in the eyes of the world. It matters whether you are healthy or not healthy, whether you are educated or not educated. It matters whether you are rich or poor—it certainly makes a difference in your life.

Yes, all these things matter, relatively speaking. But they don't matter absolutely. There is something that matters more than any of those things and that is finding the essence of who your are beyond that short-lived entity, that short-lived personalized sense of self.

You find peace not by rearranging the circumstances of your life but by realizing who you are at the deepest level.

* * *

All the misery on the planet arises due to a personalized sense of "me" or "us". That covers up the essence of who you are. When you are unaware of that inner essence, in the end, you always create misery. It's as simple as that. When you don't know who you are, you create a mind-made self as a substitute for your beautiful, divine being and cling to that fearful and needy self. Protecting and enhancing that false sense of self then becomes your primary motivating force.

* * *

Many expressions that are in common usage and sometimes the structure of language itself, reveal the fact that people don't know who they are. You say, "He lost his life." Or, "my life" as if life were something that you can possess or lose.

The truth is: you don't "have" a life, you "are" life. The one life, the one conscious that pervades the entire universe and takes temporary form to experience itself as a stone or a blade of grass, as an animal, a person, a star or a galaxy. Can you sense deep within that you already know that. Can you sense that you already are That?

* * *

For most things in life, you need time: to learn a new skill, build a house, become an expert, make a cup of tea.

Time is useless, however, for the most essential thing in life, the one thing that really matters, self-realization, which means knowing who you are beyond the surface self — beyond your name, your physical form, your history, your story.

You cannot find yourself in the past or future. The only place where you can find yourself in the Now.

The seekers look for self realization or enlightenment in the future. To be a seeker implies that you need the future. If this is what you believe, it becomes true for you: you "will" need time until you realize that you don't need time to be who you are.

* * *

When you look at a tree, you are aware of the

tree. When you have a thought or feeling, you are aware of that thought or feeling. When you have a pleasurable or painful experience, you are aware of that experience.

These seem to be true and obvious statements. Yet if you look at them very closely, you will find that in a subtle way their very structure contains a fundamental illusion, an illusion which is unavoidable when you use language. Thought and language create an apparent duality and a separate person where there is none.

The truth is you are not somebody who is aware of the tree, the thought, feeling or experience. You are the awareness or consciousness in and by which those things appear.

As you go about your life, can you be aware of yourself as the awareness in which the entire content of your life unfolds?

* * *

You say, "I want to know myself." You "are" the "I". You "are" the knowing. You are the consciousness through which everything is known and that cannot "know" itself. It "is" itself. There is nothing to know beyond that. And yet all knowing arises out of it. The "I" cannot make itself into an object of knowledge, of consciousness.

So you cannot become an object to yourself. That is the very reason the illusion of egoic identity arose— because mentally you made yourself into an object. "That's me," you say, and then you begin to have a relationship with yourself and tell others and yourself your story.

* * *

By knowing yourself as the awareness in which phenomenal existence happens, you become free of dependency on phenomena and free of self-seeking in situations, places, and conditions. In other words, what happens or doesn't happen is not that important anymore. Things lose their heaviness, their seriousness. A playfulness comes into your life. You recognize this world as a cosmic dance, the dance of form. No more and no less.

* * *

When you know who you truly are, there is an abiding alive sense of peace. You could call it joy because that's what joy is: vibrantly alive peace. It is the joy of knowing yourself as the very life essence before life takes on form. That is the joy of Being—

of being who you truly are.

* * *

Just as water can be solid, liquid, or gaseous, consciousness can be seen to be frozen as physical matter, "liquid" as mind and thought, or formless as pure consciousness. Pure consciousness is Life before it comes into manifestation and that Life looks at the world of form through "your" eyes because consciousness is who you are. When you know yourself as That, then you recognize yourself in everything. It is a state of complete clarity of perception. You are no longer an entity with a heavy past that becomes a screen of concepts through which every experience is interpreted. When you perceive without interpretation, you can then sense what it is that is perceiving. The most we can say in language is that there is a field of alert stillness in which the perception happens. Through "you", formless consciousness has become aware of itself.

* * *

Most people's lives are run by desire and fear. Desire is the need to "add" something to yourself in order to "be" yourself more fully. All fear is the fear of "losing" something, and thereby becoming diminished and being less. These two movements obscure the fact that being cannot be given or taken away. Being in its fullness is already within you, Now.

Chapter 6   Do One Thing at a Time

Doing one thing at a time means
being totally committed to the thing
that you are doing and completely
concentrating on it.

Whenever you are able, have a "look" inside yourself to see whether you are unconsciously creating conflict between the inner and the outer, between your external circumstances at that moment—where you are, who you are with, or what you are doing—and your thoughts and feelings. Can you feel how painful it is to internally stand in opposition to what "is"?

When you recognize this, you also realize that you are now free to give up this futile conflict, this inner state of war.

* * *

How often each day, if you were to verbalize your inner reality at that moment, would you have to say, "I don't want to be where I am?" What does it feel like when you don't want to be where you are—the traffic jam, your place of work, the airport lounge, the people you are with?

It is true, of course, that some places are good places to walk out of—and sometimes that may well be the most appropriate thing for you to do. In many other cases, however, walking out is not an option. In all those cases, the "I don't want to be here" is not only useless but also dysfunctional. It makes you and others unhappy.

It has been said: wherever you go, there you are. In other words: you are here. Always. Is it so hard to accept that?

* * *

Do you really need to mentally label every

sense perception and experience? Do you really need to have a reactive like/dislike relationship with life where you are in almost continuous conflict with situations and people? Or is that just a deep-seated mental habit that can be broken? Not by doing anything, but by allowing this moment to be as it is.

* * *

The habitual and reactive "no" strengthens the ego. "Yes" weakens it. Your form identity, the ego, cannot survive surrender.

* * *

"I have so much to do." Yes, but what is the quality of your doing? Driving to work, speaking to clients, working on the computer, running errands, dealing with the countless things that make up your daily life—how total are you in what you do? Is your doing surrendered or non-surrendered? This is what determines your success in life, not how much effort you make. Effort implies stress and strain, "needing" to reach a certain point in the future or accomplish a certain result.

Can you detect even the slightest element within yourself of "not wanting" to be doing what you are doing? That is a denial of life, and so a truly successful outcome is not possible.

If you can detect this within yourself, can you also drop it and be total in what you do?

* * *

"Doing one thing at a time." This is how one

Zen Master defined the essence of Zen.

Doing one thing at a time means to be total in what you do, to give it your complete attention. This is surrendered action—empowered action.

* * *

Your acceptance of what is takes you to a deeper level where your inner state as well as your sense of self no longer depend on the mind's judgment of "good" or "bad".

When you say "yes" to the "isness" of life, when you accept this moment as it is, you can feel a sense of spaciousness within you that is deeply peaceful.

On the surface, you may still be happy when it's sunny and not so happy when it's rainy; you may be happy at winning a million dollars and unhappy at losing all your possessions. Neither happiness nor unhappiness, however, go all that deep anymore. They are ripples on the surface of your Being. The background peace within you remains undisturbed regardless of the nature of the outside condition.

The "yes" to what is reveals a dimension of depth within you that is dependent neither on external conditions nor on the internal conditions of constantly fluctuating thoughts and emotions.

* * *

Surrender becomes so much easier when you realize the fleeting nature of all experiences and that the world cannot give you anything of lasting value. You then continue to meet people, to be involved in experiences and activities, but without the wants and fears of the egoic self. That is to say, you no longer demand that a situation, person, place, or event should satisfy you or make you happy. Its passing and imperfect nature is allowed to be.

And the miracle is that when you are no longer placing an impossible demand on it, every situation, person, place, or event becomes not only satisfying but also more harmonious, more peaceful.

* * *

When you completely accept this moment, when you no longer argue with what "is", the compulsion to think lessens and is replaced by an alert stillness. You are fully conscious, yet the mind is not labeling this moment in any way. This state of inner nonresistance opens you to the unconditioned consciousness that is infinitely greater than the human mind. This vast intelligence can then express itself through you and assist you, both from within and from without. That is why, by letting go of inner resistance, you often find circumstances change for the better.

* * *

Am I saying, "Enjoy this moment. Be happy?" No.

Allow the "suchness" of this moment. That's enough.

* * *

Surrender is surrender to "this moment", not to a story through which you "interpret" this moment and then try to resign yourself to it.

For instance, you may have a disability and can't walk anymore. The condition is as it is.

Perhaps your mind is now creating a story that says, "This is what my life has come to. I have ended up in a wheelchair. Life has treated me harshly and unfairly. I don't deserve this."

Can you accept the "isness of" this moment and not confuse it with a story the mind has created around it?

* * *

Surrender comes when you no longer ask, "Why is this happening to me?"

* * *

Even within the seemingly most unacceptable and painful situation is concealed a deeper good, and within every disaster is contained the seed of grace.

* * *

There are situations where all answers and explanations fail. Life does not make sense anymore. Or someone in distress comes to you for help, and you don't know what to do or say.

When you fully accept that you don't know, you give up struggling to find answers with the limited thinking mind, and that is when a greater intelligence can operate through you. And even thought can then benefit from that, since the greater intelligence can flow into it and inspire it.

Sometimes surrender means giving up trying to understand and becoming comfortable with not knowing.

* * *

Do you know of someone whose main function in life seems to be to make themselves and others miserable, to spread unhappiness? The role they play represents an intensification of the nightmare of egoic consciousness, the state of non-surrender.

* * *

Surrender, one could say, is the inner transition from resistance to acceptance, from "no" to "yes". When you surrender, your sense of self shifts from being identified with a reaction or mental judgment to being the "space around" the reaction or judgment. It is a shift from identification with form—the thought or the emotion—to being and recognizing yourself as that which has no form—spacious awareness.

* * *

Whatever you accept completely will take you to peace, including the acceptance that you cannot accept, that you are in resistance.

* * *

Leave Life alone. Let it be.

Chapter 7 Nature Education Is Free

Noticing a flower for how it
follows its natural rhythms.

We depend on nature not only for our physical survival. We also need nature to show us the way home, the way out of the prison of our own minds. We got lost in doing, thinking, remembering, anticipating—lost in a maze of complexity and a world of problems.

We have forgotten what rocks, plants, and animals still know. We have forgotten how to "be"— to be still, to be ourselves, to be where life is: Here and Now.

* * *

Whenever you bring your attention to anything natural, anything that has come into existence without human intervention, you step out of the prison of conceptualized thinking and, to some extent, participate in the state of connectedness with Being in which everything natural still exists.

To bring your attention to a stone, a tree, or an animal does not mean to "think" about it, but simply to perceive it, to hold it in your awareness.

Something of its essence then transmits itself to you. You can sense how still it is, and in doing so the same stillness arises within you. You sense how deeply it rests in Being—completely at one with what it is and where it is. In realizing this, you too come to a place of rest deep within yourself.

* * *

When walking or resting in nature, honor that realm by being there fully. Be still. Look. Listen. See how every animal and every plant is completely itself. Unlike humans, they have not split themselves in two. They do not live through mental images of themselves, so they do not need to be concerned with trying to protect and enhance those images. The deer "is" itself. The daffodil "is" itself.

All things in nature are not only one with themselves but also one with the totality. They haven't removed themselves from the fabric of the whole by claiming a separate existence: "me" and the rest of the universe.

The contemplation of nature can free you of that "me", the great troublemaker.

* * *

Bring awareness to the many subtle sounds of nature—the rustling of leaves in the wind, raindrops falling, the humming of an insect, the first birdsong at dawn. Give yourself completely to the act of listening. Beyond the sounds there is something greater: a sacredness that cannot be understood through thought.

You didn't create your body, nor are you able to control the body's functions. An intelligence greater than the human mind is at work. It is the same intelligence that sustains all of nature. You cannot get any closer to that intelligence than by being aware of your own inner energy field—by feeling the aliveness, the animating presence within the body.

* * *

The playfulness and joy of a dog, its unconditional love and readiness to celebrate life at any moment often contrast sharply with the inner state of the dog's owner—depressed, anxious, burdened by problems, lost in thought, not present in the only place and only time there is: Here and Now. One wonders: living with this person, how does the dog manage to remain so sane, so joyous?

* * *

When you perceive nature only through the mind, through thinking, you cannot sense its aliveness, its beingness. You see the form only and are unaware of the life within the form—the sacred mystery. Nature is not a commodity to be used in the pursuit of profit or knowledge or some other utilitarian purpose. The ancient forest becomes timber, the bird a research project, the mountain something to be mined or conquered.

When you perceive nature, let there be spaces of no thought, no mind. When you approach nature in this way, it will respond to you and participate in the evolution of human and planetary consciousness.

* * *

Notice how present a flower is, how surrendered to life.

* * *

The plant that you have in your home—have you ever truly looked at it? Have you allowed that familiar yet mysterious being we call "plant" to teach you its secrets? Have you noticed how deeply peaceful it is? How it is surrounded by a field of stillness? The moment you become aware of a plant's emanation of stillness and peace, that plant becomes your teacher.

* * *

Watch an animal, a flower, a tree, and see how it rests in Being. It "is" itself. It has enormous dignity, innocence, and holiness. However, for you to see that, you need to go beyond the mental habit of naming and labeling. The moment you look beyond mental labels, you feel that ineffable dimension of nature that cannot be perceived through the senses. It is a harmony, a sacredness that permeates not only the whole of nature but is also within you.

* * *

The air that you breathe is nature, as is the breathing process itself.

Bring your attention to your breathing and realize that you are not doing it. It is the breath of nature. If you had to remember to breathe, you would soon die, and if you tried to stop breathing, nature would prevail.

You reconnect with nature in the most intimate and powerful way by becoming aware of your breathing and learning to hold your attention there. This is a healing and deeply empowering thing to do. It brings about a shift in consciousness from the conceptual world of thought to the inner realm of unconditioned consciousness.

* * *

You need nature as your teacher to help you reconnect with Being. But not only do you need nature, it also needs you.

You are not separate from nature. We are all part of the One Life that manifests itself in countless forms throughout the universe, forms that are all completely interconnected. When you recognize the sacredness, the beauty, the incredible stillness and dignity in which a flower or a tree exists, you add something to the flower or the tree. Through your recognition, your awareness, nature too comes to know itself. It comes to know its own beauty and sacredness through you!

* * *

A great silent space holds all of nature in its embrace. It also holds you.

* * *

Only when you are still inside do you have access to the realm of stillness that rocks, plants, and animals inhabit. Only when your noisy mind subsides can you connect with nature at a deep level and go beyond the sense of separation created by excessive thinking.

Thinking is a stage in the evolution of life. Nature exists in innocent stillness that is prior to the arising of thought. The tree, the flower, the bird, the rock are unaware of their own beauty and sacredness. When human beings become still, they go beyond thought. There is an added dimension of knowing, of awareness, in the stillness that is beyond thought.

Nature can bring you to stillness. That is its gift to you. When you perceive and join with nature in the field of stillness, that field becomes permeated with your awareness. That is your gift to nature.

Through you nature becomes aware of itself. Nature has been waiting for you, as it were, for millions of years.

Chapter 8 Hell or Heaven

Your way of being is one of the best presents that you can give to the people around you.

How quick we are to form an opinion of a person, to come to a conclusion about them. It is satisfying to the egoic mind to label another human being, to give them a conceptual identity, to pronounce righteous judgment upon them.

Every human being has been conditioned to think and behave in certain ways—conditioned genetically as well as by their childhood experiences and their cultural environment. That is not who they are, but that is who they appear to be.

When you pronounce judgment upon someone, you confuse those conditioned mind patterns with who they are. To do that is in itself a deeply conditioned and unconscious pattern. You give them a conceptual identity, and that false identity becomes a prison not only for the other person but also for yourself.

To let go of judgment does not mean that you don't see what they do. It means that you recognize their behavior as a form of conditioning, and you see it and accept it as that. You don't construct an identity out of it for that person.

That liberates you as well as the other person from identification with conditioning, with form, with mind. The ego then no longer runs your relationships.

* * *

As long as the ego runs your life, most of your thoughts, emotions, and actions arise from desire and fear. In relationships you then either want or fear something from the other person.

What you want from them may be pleasure or material gain, recognition, praise or attention, or a strengthening of your sense of self through comparison and through establishing that you are, have, or know more than they. What you fear is that the opposite may be the case, and they may diminish your sense of self in some way.

When you make the present moment the focal point of your attention—instead of using it as a means to an end—you go beyond the ego and beyond the unconscious compulsion to use people as a means to an end, the end being self-enhancement at the cost of others. When you give your fullest attention to whoever you are interacting with, you take past and future out of the relationship, except for practical matters. When you are fully present with everyone you meet, you relinquish the conceptual identity you made for them—your interpretation of who they are and what they did in the past—and are able to interact without the egoic movements of desire and fear. Attention, which is alert stillness, is the key.

How wonderful to go beyond wanting and fearing in your relationships. Love does not want or fear anything.

* * *

To know another human being in their essence, you don't really need to know anything "about" them—their past, their history, their story. We confuse knowing about with a deeper knowing which is non-conceptual. Knowing "about" and knowing are totally different modalities. One is concerned with form, the other with the formless. One operates through thought, the other through stillness.

Knowing "about" is helpful for practical

purposes. On that level, we cannot do without it. When it is the predominant modality in relationships, however, it becomes very limiting, even destructive. Thoughts and concepts create an artificial barrier, a separation between human beings. Your interactions are then not rooted in Being, but become mind-based. Without the conceptual barriers, love is naturally present in all human interactions.

* * *

Most human interactions are confined to the exchange of words—the realm of thought. It is essential to bring some stillness, particularly into your close relationships.

No relationship can thrive without the sense of spaciousness that comes with stillness. Meditate or spend silent time in nature together. When going for a walk or sitting in the car or at home, become comfortable with being in stillness together. Stillness cannot and need not be created. Just be receptive to the stillness that is already there, but is usually obscured by mental noise.

If spacious stillness is missing, the relationship will be dominated by the mind and can easily be taken over by problems and conflict. If stillness is there, it can contain anything.

* * *

True listening is another way of bringing stillness into the relationship. When you truly listen to someone, the dimension of stillness arises and becomes an essential part of the relationship. But true listening is a rare skill. Usually, the greater part of a person's attention is taken up by their thinking. At best, they may be evaluating your words or preparing the next thing to say. Or they may not be listening at all, lost in their own thoughts.

True listening goes far beyond auditory perception. It is the arising of alert attention, a space of presence in which the words are being received. The words now become secondary. They may be meaningful or they may not make sense. Far more important than "what" you are listening to is the act of listening itself, the space of conscious presence that arises as you listen. That space is a unifying field of awareness in which you meet the other person without the separative barriers created by conceptual thinking. And now the other person is no longer "other". In that space, you are joined together as one awareness, one consciousness.

* * *

Do you experience frequent and repetitive drama
in your close relationships? Do relatively insignificant
disagreements often trigger violent arguments and
emotional pain? At the root of it lie the basic egoic
patterns: the need to be right, and, of course, for
someone else to be wrong, that is to say identification
with mental positions. There is also the ego's need to
be periodically in conflict with something or someone
in order to strengthen its sense of separation between
"me" and the "other", without which it cannot
survive.

In addition, there is the accumulated emotional
pain from the past that you and each human being
carries within, both from your personal past as well
as the collective pain of humanity that goes back a
long, long time. This "pain-body" is an energy field

within you that sporadically takes you over because it needs to experience more emotional pain for itself to feed on and replenish itself. It will try to control your thinking and make it deeply negative. It loves your negative thoughts since it resonates with every frequency and so can feed on them. It will also provoke negative emotional reactions in people close to you, especially your partner, in order to feed on the ensuing drama and emotional pain.

How can you free yourself from this deep-seated unconscious identification with pain that creates so much misery in life? Become aware of it. Realize that it is not who you are and recognize it for what it is, past pain. Witness it as it happens in your partner or in yourself. When your unconscious identification with it is broken, when you are able to observe it within yourself, you don't feed it anymore and it will gradually lose its energy charge.

* * *

Human interaction can be hell. Or it can be a great practice.

* * *

When you look upon another human being and feel great love towards them, or when you contemplate beauty in nature and something within you responds deeply to it, close your eyes for a moment and feel the essence of that love or that beauty within you, inseparable from who you are, your true nature. The outer form is a temporary reflection of what you are within, in your essence. That is why love and beauty can never leave you, although all outer forms will.

* * *

What is your relationship with the world of objects, the countless things that surround you, and that you handle everyday. The chair you sit on, the pen, the car, the cup. Are they to you merely a means to an end or do you occasionally acknowledge their existence, their being, no matter how briefly, by noticing them and giving them your attention? When you get attached to objects, when you are using them to enhance your worth in your own eyes and in the eyes of others, concern about things can easily take over your whole life.

When there is self-identification with things, you don't appreciate them for what they are because you are looking for yourself in them.

When you appreciate an object for what it is, when you acknowledge its being without mental projection, you cannot "not" feel grateful for its existence. You may also sense that it is not really inanimate, that it only appears so to the senses. Physicist will confirm that on a molecular level, it is indeed, a pulsating energy field.

Through selfless appreciation of the realm of things, the world around you will begin to come alive for you in ways you cannot even begin to comprehend with the mind.

* * *

Whenever you meet anyone, no matter how briefly, do you acknowledge their being by giving them your full attention, or are you reducing them to a means to an end, a mere function or role? What is the quality of your relationship with the cashier at the supermarket, the parking attendant, the repair man, the "customer"?

A moment of attention is enough. As you look at them or listen to them, there is an alert stillness—perhaps only two or three seconds. Perhaps longer. That is enough for something more real to emerge than the roles we usually play and identify with.

All roles are part of the conditioned consciousness that is the human mind. That which emerges through the act of attention is the unconditioned—who you are in your essence underneath your name and form. You are no longer acting out a script. You become real. When that dimension emerges from within you, it also draws it forth from within the other person.

Ultimately, of course, there is no other and you are always meeting yourself.

Chapter 9   Take It Easy

If you can learn to accept or even welcome these losses in your life, maybe you will find the emptiness and initial anxiousness will become profound inner peace.

When you walk though a forest that has not been tamed and interfered with by man, you will see not only abundant life all around you, but you will also encounter fallen trees and decaying trunks, rotting leaves and decomposing matter at every step. Wherever you look, you will find death as well as life.

Upon closer scrutiny, however, you will discover that the decomposing tree trunk and rotting leaves not only give birth to new life, but are full of life themselves. Microorganisms are at work. Molecules are rearranging themselves. So death isn't to be found anywhere. There is only the metamorphosis of life forms. What can you learn from this?

Death is not the opposite of life. Life has no

opposite. The opposite of death is birth. Life is eternal.

* * *

Sages and poets throughout the ages have recognized the dreamlike quality of human existence—seemingly so solid and real and yet so fleeting that it could dissolve at any moment.

At the hour of your death, the story of your life may, indeed, appear to you like a dream that is coming to an end. Yet even in a dream there must be an essence that is real. There must be a consciousness in which the dream happens; otherwise, it would not be.

That consciousness—does the body create it or does consciousness create the dream of body, the dream of somebody?

Why did most of those who went through a near-death experience lost their fear of death? Reflect upon this.

* * *

Of course you know you are going to die, but that remains a mere mental concept until you meet death "in person" for the first time: through a serious illness or an accident that happens to you or someone close to you, or through the passing away of a loved one, death enters your life as the awareness of your own mortality.

Most people turn away from it in fear, but if you do not flinch and face the fact that your body is fleeting and could dissolve at any moment, there is some degree of disidentification, however slight, from your own physical and psychological form, the "me".

When you see and accept the impermanent nature of all life forms, a strange sense of peace comes upon you.

Through facing death, your consciousness is freed to some extent from identification with form. This is why in some Buddhist traditions, the monks regularly visit the morgue to sit and meditate among the dead bodies.

There is still a widespread denial of death in Western culture. Even old people try not to speak or think about it, and dead bodies are hidden away. A culture that denies death inevitably becomes shallow and superficial, concerned only with the external form of things. When death is denied, life loses its depth. The possibility of knowing who we are beyond name and form, the dimension of the transcendent, disappears from our lives because death is the opening into that dimension.

* * *

People tend to be uncomfortable with endings, because every ending is a little death. That's why in many languages, the word for "good-bye" means "see you again".

Whenever an experience comes to an end—a gathering of friends, a vacation, your children leaving home—you die a little death. A "form" that appeared in your consciousness as that experience dissolves. Often this leaves behind a feeling of emptiness that most people try hard not to feel, not to face.

If you can learn to accept and even welcome the endings in your life, you may find that the feeling of emptiness that initially felt uncomfortable turns into a sense of inner spaciousness that is deeply peaceful.

By learning to die daily in this way, you open

yourself to Life.

* * *

Most people feel that their identity, their sense of self, is something incredibly precious that they don't want to lose. That is why they have such fear of death.

It seems unimaginable and frightening that "I" could cease to exist. But you confuse that precious "I" with your name and form and a story associated with it. That "I" is no more than a temporary formation in the field of consciousness.

As long as that form identity is all you know, you are not aware that this preciousness is your own essence, your innermost sense of "I Am", which is consciousness itself. It is the eternal in you—and that's the only thing you cannot lose.

* * *

Whenever any kind of deep loss occurs in your life—such as loss of possessions, your home, a close relationship; or loss of your reputation, job, or physical abilities—something inside you dies. You feel diminished in your sense of who you are. There may also be a certain disorientation. "Without this...who am I?"

When a form that you had unconsciously identified with as part of yourself leaves you or dissolves, that can be extremely painful. It leaves a hole, so to speak, in the fabric of your existence.

When this happens, don't deny or ignore the pain or the sadness that you feel. Accept that it is there. Beware of your mind's tendency to construct a story around that loss in which you are assigned the role of victim. Fear, anger, resentment, or self-pity are the emotions that go with that role. Then become aware of what lies behind those emotions as well as behind the mind-made story: that hole, that empty space. Can you face and accept that strange sense of emptiness? If you do, you may find that it is no longer a fearful place. You may be surprised to find peace emanating from it.

* * *

How short-lived every human experience is, how fleeting our lives. Is there anything that is not subject to birth and death, anything that is eternal?

Consider this: if there were only one color, let us say blue, and the entire world and everything in it were blue, then there would be no blue. There needs to be something that is not blue so that blue can be recognized; otherwise, it would not "stand out", would not exist.

In the same way, does it not require something that is not fleeting and impermanent for the fleetingness of all things to be recognized? In other words: if everything, including yourself, were impermanent, would you even know it? Does the fact that you are aware of and can witness the short-lived nature of all forms, including your own, not mean that there is something in you that is not subject to decay?

When you are twenty, you are aware of your body as strong and vigorous; sixty years later, you are aware of your body as weakened and old. Your thinking too may have changed from when you were twenty, but the awareness that knows that your body is young or old or that your thinking has changed has undergone no change. That awareness is the eternal in you—consciousness itself. It is the formless One Life. Can you lose It? No, because you are It.

* * *

Some people become deeply peaceful and almost luminous just before they die, as if something is shining through the dissolving form.

Sometimes it happens that very ill or old people become almost transparent, so to speak, in the last few weeks, months, or even years of their lives. As they look at you, you may see a light shining through their eyes. There is no psychological suffering left. They have surrendered and so the person, the mind-made egoic "me", has already dissolved. They have "died before they died" and found the deep inner peace that is the realization of the deathless within themselves.

* * *

To every accident and disaster there is a

potentially redemptive dimension that we are usually unaware of.

The tremendous shock of totally unexpected, imminent death can have the effect of forcing your consciousness completely out of identification with form. In the last few moments before physical death, and as you die, you then experience yourself as consciousness free of form. Suddenly, there is no more fear, just peace and a knowing that "all is well" and that death is only a form dissolving. Death is then recognized as ultimately illusory—as illusory as the form you had identified with as yourself.

* * *

Death is not an anomaly or the most dreadful of all events as modern culture would have you believe, but the most natural thing in the world, inseparable from and just as natural as its other polarity—birth.

Remind yourself of this when you sit with a dying person.

When you sit with a dying person, do not deny any aspect of that experience. Do not deny what is happening and do not deny your feelings. The recognition that there is nothing you can do may make you feel helpless, sad, or angry. Accept what you feel. Then go one step further: accept that there is nothing you can do, and accept it completely. You are not in control. Deeply surrender to every aspect of that experience, your feelings as well as any pain or discomfort the dying person may be experiencing. Your surrendered state of consciousness and the stillness that comes with it will greatly assist the dying person and ease their transition. If words are called for, they will come out of the stillness within you. But they will be secondary.

With the stillness comes the benediction: peace.

Chapter 10 Survive Adversity

Certain pain and trouble are almost necessities of life. And the unfortunate can be found everywhere while people being worth of envy are pretty rare. So what we can do is to accept the life that we are living, do not trouble ourselves and try to make it more interesting and richer. Maybe then we will come across our "a-ha" moment.

The interconnectedness of all things: Physicists now confirm it. Nothing that happens is an isolated event. It only appears to be. The more we judge and label it, the more we isolate it.

The wholeness of life becomes fragmented through our thinking. Yet the totality of life has brought this event about. It is part of the web of interconnectedness that is the cosmos. This means whatever is could not be otherwise.

In most cases, we cannot begin to understand what role a seemingly senseless event may have within the totality of the cosmos but recognizing its inevitability within the vastness of the whole can be the beginning of an inner acceptance of what is and thus a realignment with the wholeness of life.

* * *

True freedom and the end of suffering is living in such a way as if you had completely chosen whatever you feel or experience at this moment. This inner alignment with Now is the end of suffering.

* * *

Is suffering really necessary. Yes and no. If you had not suffered as you have, there would be no depths to you as a human being—no humility, no compassion. You would not be listening to this now. Suffering cracks open the shell of ego. And then comes a point where it has served its purpose.

Suffering is necessary until you realize that it is unnecessary.

* * *

Unhappiness needs a mind-made "me" with a story, the conceptual identity. It needs time—past and future. When you remove time from your unhappiness, what is it that remains? The "suchness" of this moment remains. It may be a feeling of heaviness of heaviness, agitation, tightness, anger or even nausea. That is not unhappiness and it is not a personal problem.

There is nothing personal in human pain. It is simply an intense pressure or an intense energy you feel somewhere in the body. By giving it attention, the feeling doesn't turn into thinking and thus reactivate the unhappy "me". See what happens when you just allow a feeling to be.

* * *

Much suffering, much unhappiness arises when you take each thought that comes into your head for the truth. Situations don't make you unhappy. They may cause you physical pain, but they don't make you unhappy. Your thoughts make you unhappy. Your interpretations, the stories you tell yourself make you unhappy. "The thoughts I'm thinking right now are making me unhappy." This realization breaks you unconscious identification with those thoughts.

* * *

"What a miserable day."

"He didn't have the decency to return my call."

"She let me down."

Little stores we tell ourselves and others, often in the form of complaints. They are unconsciously designed to enhance our always deficient sense of self through being "right" and making something or someone "wrong". Being "right" places us in a position of imagined superiority and so strengthens our false sense of self, the ego. This also creates some kind of enemy: yes, the ego needs enemies to define its boundary. And even the weather can serve that function. Through habitual mental judgment and emotional contraction you have a personalized reactive relationship to people and events in your life. These are all forms of self-created suffering but are not recognized as such because to the ego they are satisfying. The ego enhances itself through reactivity and conflict. How simple life would be without those stories.

"It is raining."

"He did not call."

"I was there, she was not."

* * *

When you are suffering, when you are unhappy, stay totally with what is Now. Unhappiness or problems cannot survive in the Now.

* * *

Suffering begins when you mentally name or label a situation in some way as undesirable or bad. You resent a situation and that resentment personalizes it and brings in the reactive "me". Naming and labeling are habitual but that habit can be broken. Start practicing "not naming" with small things. If you miss the plane, drop and break a cup, or slip and fall in the mud, can you refrain from naming the experience as bad or painful? Can you immediately accept the "isness" of that moment?

Naming something as bad causes an emotional contraction within you. When you let it be without naming it, enormous power is suddenly available to you. The contraction cuts you off from that power, the power of life itself.

* * *

They ate the fruit of the tree of the knowledge of good and evil. Go beyond good and bad by refraining from mentally labeling anything as good or bad. When you go beyond the habitual naming, the power of the universe moves through you. When you are in a nonreactive relationship to experiences, what you would have called "bad" before often turns around quickly, if not immediately through the power of life itself. Watch what happens when you don't name an experience as "bad" and instead bring an inner acceptance, an inner "yes" to it, and so let it be as it is.

* * *

Whatever your life situation is, how would you feel if you completely accepted it as it is— right Now?

* * *

There are many subtle and not so subtle forms of suffering that are so normal, they are usually not recognized as suffering and may even feel satisfying to the ego—irritation, impatience, anger, having an issue with something or someone, resentment, complaining. You can learn to recognize all those forms of suffering as they happen and know at this moment I am creating suffering for myself.

If you are in the habit of creating suffering for yourself, you are probably creating suffering for others too. These unconscious mind patters tend to come to an end simply by making them conscious, by becoming aware of them as they happen. You cannot be conscious "and" create suffering for yourself.

* * *

This is the miracle; Behind every condition, person or situation that appears "bad" or "evil", lies concealed a deeper good. That deeper good reveals itself to you—both within and without—through inner acceptance of what "is". "Resist not evil" is one of the highest truths of humanity.

* * *

A dialog:

"Accept what is."

"I truly cannot. I am agitated and angry about this."

"Then accept what is."

"Accept that I am agitated and angry? Accept that I cannot accept?"

"Yes. Bring acceptance into your non-acceptance. Bring surrender into your non-surrender. Then see what happens."

* * *

Chronic physical pain is one of the harshest

teachers you can have. "Resistance is futile" is its teaching. Nothing could be more normal than an unwillingness to suffer. Yet, if you can let go of that unwillingness, and instead allow the pain to be there, you may notice a subtle inner separation from the pain, a space between you and the pain, as it were. This means to suffer consciously, willingly. When you suffer consciously, physical pain can quickly burn up the ego in you, since ego consists largely of resistance. The same is true of extreme physical disability.

图书在版编目（CIP）数据

人生不必太用力 / (德) 埃克哈特·托利著；李密
译. —— 南京：江苏凤凰文艺出版社, 2021.8（2025.9重印）
书名原文: Stillness Speaks
ISBN 978-7-5594-6160-5

Ⅰ.①人… Ⅱ.①埃…②李… Ⅲ.①人生哲学 – 通
俗读物 Ⅳ.①B821-49

中国版本图书馆CIP数据核字(2021)第141616号

著作权合同登记号：10-2017-096

人生不必太用力

（德）埃克哈特·托利 著　李 密 译

责任编辑　李龙姣

装帧设计　一伙设计

出版发行　江苏凤凰文艺出版社

　　　　　南京市中央路 165 号，邮编：210009

网　　址　http://www.jswenyi.com

印　　刷　唐山富达印务有限公司

开　　本　787 毫米 × 1092 毫米　1/32

印　　张　7.5

字　　数　60 千字

版　　次　2021 年 8 月第 1 版

印　　次　2025 年 9 月第 5 次印刷

书　　号　ISBN 978-7-5594-6160-5

定　　价　45.00 元